Hybrid Mobile Development with Ionic

Build high performance hybrid applications with HTML, CSS, and JavaScript

Gaurav Saini

BIRMINGHAM - MUMBAI

Hybrid Mobile Development with Ionic

First published: April 2017

Production reference: 1250417

Published by Packt Publishing Ltd.
Livery Place
35 Livery Street
Birmingham
B3 2PB, UK.
ISBN 978-1-78528-605-6

www.packtpub.com

Author

Gaurav Saini

Reviewers

Nitish Sinha

Commissioning Editor

Dipika Gaonkar

Acquisition Editor

Reshma Raman

Content Development Editor

Trusha Shriyan

Technical Editor

Varsha Shivhare

Copy Editor

Safis Editing

Project Coordinator

Kinjal Bari

Proofreader

Safis Editing

Indexer

Mariammal Chettiyar

Graphics

Kirk D'Penha

Production Coordinator

Deepika Naik

Open-source enthusiastic, Javascript full stack developer with Node.js, AngularJS, electron and Ionic. Currently working on IOT projects with BLE Beacons. Gaurav have been researching on VOIP technologies, IOT and material design whenever he get time from his daily activities. Gaurav is also associated with Apache software foundation where he is contributor and committer to Apache Roller, Apache Fineract and Apache OFBiz projects. He was selected and invited as speaker at multiple conferences Mifos Techdays, ApacheCON, Mifos Summit and also into many local colleges/institutions.

Gaurav was a student intern Google Summer of Code Student in 2014 and Mentored in 2016 with Mifos Initiative worked on AngularJs based community-app. He was selected as delegate for GSOC Summit in 2014 and 2016 San Fransico. He had worked on Google Adwords and analytics in past and was semi-finalist in Google Online Marketing Challenge in 2012.

I would like to thank my parents and my late uncle Raj, without them it would be really difficult for me to reach to here and write a book. Also, at difficult times my best friend Seema came with full support to continue working hard.

A special thanks also goes to my interns and content editor as they have been helping alot during the course of the book. It was pleasure to work with great minds around you.

Lastly, I want to thank you all for showing interest in this book and I hope that I can pass my enthusiasm to join Ionic community and take your exciting journey to new heights.

Nitish Sinha is a techie from Jharkhand, and is believer of Steve Jobs words:

The only way to do great work is to love what you do.

Nitish have managed to complete various projects as a freelancer and had the good fortune to be able to channel this obsession into something productive.

He joined TERI, India Habitat Centre, New Delhi as software developer, dealing with Angular.js and Ionic based hybrid apps along with PHP based web applications.

I would like to thanks all my mentors at each and every stage, whose golden words are inspiration for me.

For support files and downloads related to your book, please visit www.PacktPub.com.

Did you know that Packt offers eBook versions of every book published, with PDF and ePub files available? You can upgrade to the eBook version at www.PacktPub.com and as a print book customer, you are entitled to a discount on the eBook copy. Get in touch with us at service@packtpub.com for more details.

At www.PacktPub.com, you can also read a collection of free technical articles, sign up for a range of free newsletters and receive exclusive discounts and offers on Packt books and eBooks.

https://www.packtpub.com/mapt

Get the most in-demand software skills with Mapt. Mapt gives you full access to all Packt books and video courses, as well as industry-leading tools to help you plan your personal development and advance your career.

Why subscribe?

- Fully searchable across every book published by Packt
- Copy and paste, print, and bookmark content
- On demand and accessible via a web browser

Thanks for purchasing this Packt book. At Packt, quality is at the heart of our editorial process. To help us improve, please leave us an honest review on this book's Amazon page at https://www.amazon.com/dp/1785286056.

If you'd like to join our team of regular reviewers, you can e-mail us at customerreviews@packtpub.com. We award our regular reviewers with free eBooks and videos in exchange for their valuable feedback. Help us be relentless in improving our products!

Table of Contents

Preface

This book teaches you how to master in hybrid mobile development using Ionic. Ionic is an open source, front-end framework that allows you to develop hybrid mobile apps without any native-language hassle for each platform. It offers a library of mobile-optimized HTML, CSS, and JS components for building highly interactive mobile apps.

Initial idea to write this book is to help understand developers what is the building blocks and necessary points to look for while you are developing an enterprise level or highly scalable Ionic application. Also, in the later half of the book we will target some latest features in Ionic such as IOT and PWA support. We have made sure that every concept addressed in this book is explained in easy language, and also with a technical example.

What this book covers

Chapter 1, *Getting Started with Ionic 3*, we will be going through all the essentials of Ionic, which will help in building large scale and enterprise grade applications.

Chapter 2, *Ionic Components*, will guide you towards Ionic components and how we can customize them according to our applications.

Chapter 3, *Ionic Native and Plugins*, covers almost everything regarding Cordova plugins, Ionic Native and some common and advance plugins.

Chapter 4, *Ionic Platform and Services*, covers all the Ionic Platform services and how to use them in best way in your application.

Chapter 5, *Authentication, Authorization, and Security*, will be dealing with authentication, securing Ionic application and demonstrating authorization.

Chapter 6, *TasteBite App with Firebase*, focuses on building entire application from start to end by using Backend as a Service platform Firebase.

Chapter 7, *Ionic, IOT and Beacons*, we will be discussing about IOT, BLE and Physical Web. Besides this will demonstrate ibeacon based proximity application where your laptop will screen lock as soon as you will move away from your laptop.

Chapter 8, *Ionic + PWA = Magic*, we will be looking into Progressive Web Applications and its support with Ionic with a demonstration of offline-first currency converter application.

What you need for this book

You'll need Sublime or Visual Studio Code for code editing. Both of these are available free for download. For BLE demo you will need a laptop with Bluetooth 4 supported or Bluetooth BLE beacon.

Who this book is for

This book is for intermediate-level application developers who have some basic knowledge of Ionic. Those who want to migrate their applications from Ionic 1 to Ionic 3 this book will be very helpful and who want to get latest technology updates such as PWA support with Ionic.

Conventions

In this book, you will find a number of text styles that distinguish between different kinds of information. Here are some examples of these styles and an explanation of their meaning.

Code words in text, database table names, folder names, filenames, file extensions, pathnames, dummy URLs, user input, and Twitter handles are shown as follows: "For setting up the APIs endpoints we have created constants file where we declared constant API_URL"

A block of code is set as follows:

```
{
  "name": "ionic2-auth",
  "app_id": "",
  "v2": true,
  "proxies": [
  {
   "path": "/api",
   "proxyUrl": "http://yourwebsite.com /api"
  }
 ]
}
```

When we wish to draw your attention to a particular part of a code block, the relevant lines or items are set in bold:

```
// src/providers/constants.ts
// While deploying application we switch to real URL
```

```
export const API_URL: string = 'http://localhost:8100/api';
```

Any command-line input or output is written as follows:

```
$ chromium-browser –disable-web-security
```

New terms and **important words** are shown in bold. Words that you see on the screen, for example, in menus or dialog boxes, appear in the text like this: "**Cancel** and **OK** buttons. We have used this here for sorting the products according to relevance price or other sorting values."

 Warnings or important notes appear in a box like this.

 Tips and tricks appear like this.

Reader feedback

Feedback from our readers is always welcome. Let us know what you think about this book-what you liked or disliked. Reader feedback is important for us as it helps us develop titles that you will really get the most out of.

To send us general feedback, simply e-mail feedback@packtpub.com, and mention the book's title in the subject of your message.

If there is a topic that you have expertise in and you are interested in either writing or contributing to a book, see our author guide at www.packtpub.com/authors.

Customer support

Now that you are the proud owner of a Packt book, we have a number of things to help you to get the most from your purchase.

Downloading the example code

You can download the example code files for this book from your account at http://www.packtpub.com. If you purchased this book elsewhere, you can visit http://www.packtpub.com/support and register to have the files e-mailed directly to you.

You can download the code files by following these steps:

1. Log in or register to our website using your e-mail address and password.
2. Hover the mouse pointer on the **SUPPORT** tab at the top.
3. Click on **Code Downloads & Errata**.
4. Enter the name of the book in the **Search** box.
5. Select the book for which you're looking to download the code files.
6. Choose from the drop-down menu where you purchased this book from.
7. Click on **Code Download**.

Once the file is downloaded, please make sure that you unzip or extract the folder using the latest version of:

*   WinRAR / 7-Zip for Windows
*   Zipeg / iZip / UnRarX for Mac
*   7-Zip / PeaZip for Linux

The code bundle for the book is also hosted on GitHub at `https://github.com/PacktPubl ishing/Hybrid-Mobile-Development-with-Ionic`. We also have other code bundles from our rich catalog of books and videos available at `https://github.com/PacktPublishing/`. Check them out!

Downloading the color images of this book

We also provide you with a PDF file that has color images of the screenshots/diagrams used in this book. The color images will help you better understand the changes in the output. You can download this file from `https://www.packtpub.com/sites/default/files/downloads/HybridMobileDevelopmentw ithIonic_ColorImages.pdf`.

Errata

Although we have taken every care to ensure the accuracy of our content, mistakes do happen. If you find a mistake in one of our books-maybe a mistake in the text or the code-we would be grateful if you could report this to us. By doing so, you can save other readers from frustration and help us improve subsequent versions of this book. If you find any errata, please report them by visiting http://www.packtpub.com/submit-errata, selecting your book, clicking on the **Errata Submission Form** link, and entering the details of your errata. Once your errata are verified, your submission will be accepted and the errata will be uploaded to our website or added to any list of existing errata under the Errata section of that title.

To view the previously submitted errata, go to https://www.packtpub.com/books/content/support and enter the name of the book in the search field. The required information will appear under the **Errata** section.

Piracy

Piracy of copyrighted material on the Internet is an ongoing problem across all media. At Packt, we take the protection of our copyright and licenses very seriously. If you come across any illegal copies of our works in any form on the Internet, please provide us with the location address or website name immediately so that we can pursue a remedy.

Please contact us at copyright@packtpub.com with a link to the suspected pirated material.

We appreciate your help in protecting our authors and our ability to bring you valuable content.

Questions

If you have a problem with any aspect of this book, you can contact us at questions@packtpub.com, and we will do our best to address the problem.

1
Getting Started with Ionic 3

Ionic has been evolving a lot over the years; ever since Ionic v1 has launched there has been continuous development and improvement going on with the framework. One of the smartest moves was to launch Ionic v2, which is based on Angular 2 by the Ionic team. Recently, Ionic 3 was launched, which is based on Angular 4 and TypeScript 2.2. Here in this book we will mainly be going through Ionic 2 and 3 and their advanced concepts. We will cover topics such as building an e-commerce app with Bluetooth beacons support for malls, integration with Firebase, and PWA support with Ionic applications. We will start with support for all three platforms: iOS, Android, and Windows, and will be demonstrating applications based on Internet of Things, which will be really exciting.

In this chapter, we will be going through all the essentials of Ionic, which will help to build large scale and enterprise grade applications. We have been working on Ionic 1x for the last 2-3 years and during that journey we have seen so much improvement in the framework and multiple features developed. After Angular 2 was released, a lot more improvement came in the framework and then the Ionic team planned on developing Ionic 2, which is based on Angular 2. Ionic 3 and Angular 4 are on similar lines and do not have major framework changes such as we had when we shifted from Ionic v1 to v2. With Ionic 3 you will now see improved performance, reduced code complexity, and the ability to build scalable applications has increased. Ionic is now targeting mobile web and desktop applications, which helps developers running same codebase everywhere. Ionic Native is another example of how the Ionic team is improving the product day by day and making developers' lives easier. We will initially look into some of the basics of Angular and Ionic, also how TypeScript can be a big asset for building enterprise applications. Ionic CLI and Ionic Cloud products are as well as improving regularly, which we can use to speed up our development time and efficiently test or share our applications. In this chapter, we will cover:

- Angular 4, Ionic 3, and TypeScript
- Installation and setup
- Directory structure and modularity

- Theming with SASS
- Ionic CLI

Angular 4, Ionic 3, and TypeScript

Many of us coming from the Angular 1 world to Angular 2 or 4 will surely see that it's an entire rewrite of the application and a steep learning curve. I felt the same when I initially started Ionic and Angular, but gradually as I read about the concepts of Angular, many of the problems we used to face in Angular 1 were automatically solved without any effort. Initially you will miss the old controllers, services, filters, and other concepts in Angular 1 and most importantly the navigation and routing. But when you dive deep into the topics you will find that corresponding modules such as **component, providers,** and **pipe** are available and that they can be used in similar ways. Navigation and router are also now used as push and pop mechanisms for navigating from one page to another page. In the initial beta versions of Ionic 2, we don't have proper URL-based routing as users can't land on a specific page in the application. But after a stable Angular-router is released it has been added to Ionic 2, which has opened the path for better support for progressive web apps, which will allow the same Ionic apps to be shipped as mobile web applications. Ionic 3 also added support for responsive grids, which will help when we will be building desktop applications. Lazy Loading is another important feature added, which reduces the initial loading time of the application. Still, as Ionic 3 is in initial days, we can expect these features to become a lot more stable. As, we know Ionic 2 and 3 did not have any major framework changes, so initially we will be comparing Angular 1 and 2 in this chapter so it helps users understand the difference and how they can migrate to the latest Ionic versions.

Angular and Ionic myths

As you go forward, you will hear many myths that are not true in respect to Angular 2 or upgrading to Angular 4. One of the most common is the Angular 4 doesn't support two-way data binding, which is not true. The Angular team has made sure that we use forms as simple as we use in Angular 4, they just have a new syntax:

```
[(property-name)]="expression"

<input type="text"
    [(ngMmodel)]="model.name" />
```

We still have two-way data binding. Although, to improve the Angular digest cycle and performance by default, we don't use two-way data binding. It uses a unidirectional binding that we can extend to two-way binding when we require it. You can refer to Angular template syntax to demystify all other syntax's and their uses available with Angular 4.

Ionic lack URL navigation is another discussion going on at the Ionic forum. Again this is also partially correct; because Angular-router was not stable enough, the Ionic team decided to remove it in the initial beta version. As of now, Angular-router is stable and will be in Ionic. The Ionic team understands how critical it is to have a proper routing mechanism and to make a developer feel at home while moving towards Ionic 3.

Upgradation is entirely rewritten: Angular 2 is itself rewritten, but you don't have to rewrite your app also. Initially when I started I felt the same and thought that we had to redevelop our current application. But there are many tools and documentation (`http://ionicframework.com/files/Ionic2Migration.pdf`) available for a smooth upgrade from Ionic 1 to Ionic 2 or 3. You just have to be careful as if you have a complex application then you might have to do many things manually. But one thing is for sure that upgrading the application will surely not take that much time when you build the same application from scratch if you have proper expertise and a concept of Angular 2. One of the reasons behind this is that you already have a running application where you just have to assemble the moving parts in different ways, as most of the application logic will be same.

Personally, my learning process with Ionic didn't think too much about I upgrading to a new version. Rather I took it as new improved concepts that came in new Angular versions, designed to ease the learning and understanding.

Mapping Ionic 1 to Ionic 3

Life will be really easy for any developer who can easily understand the mapping of different concepts that were in Angular 1 and how they are
catered to in Angular 4. Also, I am sure that most of the questions will be answered automatically.

Controllers match components

Components are the backbone of Angular and Ionic applications; you will see that almost everything is a component. You can compare controllers in Angular 1 to components in Angular 4. In controllers we used to define most of our code's logical part. We used to register the controllers with our main module. Now, in a similar way, we define our code in components in Angular 4 and if required we can export that `Component` class:

```
// Angular & Ionic 1

angular.module('wedding.controllers', [])
.controller('LoginCtrl',
function($scope, CategoryService) {
// controller function and DI of CategoryService
}
);

// Angular 4 & Ionic 32

import {Component} from '@angular/core';
import {NavController} from 'ionic-angular';

@Component({
templateUrl: 'build/pages/catalog/categories.html'
})
export class CategoryPage {
// DI of NavController for navigation
  constructor(private navCtrl: NavController) {
  this.nav = navCtrl;
}
}
```

We have dependency injection in Angular 1 as `controller` function arguments, while in Angular 4 we pass it inside the `constructor` function. Many other things such as the IIFE syntax, which we have to define in Angular 1 for keeping out controller code of the global namespace, now are not required in Angular 4 because ES 2015 modules handle name spacing for us. Also, as you can see, we have exported the `CategoryPage` class, we can now import it wherever this module is required.

Another major change is that `$scope` is replaced by the this keyword. `$scope` had many performance issues and already in Angular 1 developers have reduced the usage of `$scope`.

Filters match pipes

Angular 4 pipes are similar to what we use `Filters` for in Angular 1. `Pipes` provide the same formatting and transformation for data in the template. Almost all of the inbuilt filters that Angular 1 has correspond to pipes in Angular 4:

```
// Filter - Angular 1
<td>{{movie.price | currency}}</td>

// Pipes in Angular 4
<td>{{movie.price | currency:'USD':true}}
</td>
```

Due to performance reasons, filter and `orderBy` filters have been removed from `Pipes`, although you can at anytime build a custom pipe if similar code is reused in multiple templates.

Services match providers

Services and Factory are important parts of Angular 1x where we communicate with a remote server for data. We used to define APIs call inside `factory` functions that controllers calls for fetching data from the servers:

```
// Factory method in Angular 1x

angular.module('wedding.services', [])

// DI of $http for HTTP calls to servers
// $q  for promises
.factory('CategoryService', function ($http, $q) {
    var catalog   = {};

    catalog.getCategories = function () {
        // here we will call APIs
    }
})
```

Now you will see how we have migrated our code to Angular 4 providers. The same `getCategories()` method that was inside factory in Angular 1, will now be moved inside the `CategoryData()` class in Angular 4:

```
// Provider in Angular 4

import { Injectable } from '@angular/core';
import { Http } from '@angular/http';

@Injectable()
export class CategoryData {

  constructor(private http: Http) {}

  getCategories() {

    return new Promise(resolve => {
      // We're using Angular Http provider
         to request the data,
      // then on the response it'll map the
         JSON data to a parsed JS object.
      // Next we process the data and
         resolve the promise with the new
         data.

  this.http.get('www.veloice.com/data.json').subscribe(res
     => {
        // we've got back the raw data, now
           generate the core schedule data
        // and save the data for later
           reference
        this.data = this.processData(res.json());
        resolve(this.data);
      });
    });
  }
}
```

You will notice that the `Provider` class has a `@Injectable` decorator. This decorator lets Angular 4 know that the specific class can be used with the dependency injector.

TypeScript comes to the rescue

TypeScript is basically a superset of Javascript, which is a statically typed language. TypeScript does similar to what LESS or SASS does to CSS. TypeScript compiles to Javascript and has almost the same syntax. Anyone from an object-oriented world will find it really, familiar with concepts such as classes, constructors, inheritance, and interfaces. Many developers will be confused with all these keywords such as ES5, ES6, TypeScript, AtScript, and many others. JavaScript is a community-driven language, and accordingly, browser manufacturers implement those features that are mostly used. Currently, ES5, that is ECMAScript 5, is a standardization by ECMA international. ES6 and ES7 are future standards of JavaScript with tons of new features. We can develop applications in ES6 and ES7 directly with the use of Babel, which is a transpiler. Transpilers are source-to-source compilers that take input in one programming language and output the equivalent code in another language. Although it is recommended to use TypeScript as the majority of work going on in the Ionic community is on TypeScript. Also, you will easily get resources and community help.

When we start a new Ionic 3 project it create the project with TypeScript, although in Ionic 2 and CLI v2 we had option for creating project with TypeScript. As mentioned, we should use TypeScript because, as the project will become a large scale project, things will be easier to maintain and understand. I want to clear up one important point here about Angular 4: whether it's TypeScript or ES 6 whenever you run the Ionic server, our `app` folder is transpiled into ES5, as that is what currently all major browsers support. There are many new features available in TypeScript that we will be mainly using in our Ionic or Angular applications. Some of them are listed here:

- Class and module support
- Static type-checking
- Similar syntax to other object-oriented languages
- ES6 features such as template string support
- Decorator, annotation support, and dependency injection
- Arrow functions

Installation and setup

As we are going through Ionic 3, most of the developers have previous experience of Ionic installation. We can install `ionic` via the NPM package; currently Ionic CLI v3 is in beta which has entirely changed how CLI works and now the default project that Ionic will create is based on Ionic 3:

```
$ npm install -g ionic
```

It's recommended to use the 6x version for node and the 3x version of npm. This beta release supports both Ionic 1 and 2 projects. Now let's get started with a project:

```
// will create Ionic 3x project
$ ionic start wedding-planner sidemenu
// will create Ionic 1x project
$ ionic start wedding-planner sidemenu --type ionic1
// move to projedct directory
  $ cd wedding-planner

  // run project with ionic serve
  $ ionic serve
```

Now the next steps will be to add specific platforms and plugins required in the application:

```
// For adding specific platforms
$ ionic cordova:platform add android

// running application on real device
$ ionic cordova:run android

// running inside an emulator
$ ionic cordova:emulate android
```

Next, we will be looking at how we can get started with an Ionic 2 application code base. You have to understand how each file and folder is structured and how they work.

Directory structure and modularity

Previously in Ionic 1x, we didn't have the best directory or project structure. Slowly as Ionic and Angular evolved, developers tended to move towards the modular approach for organizing their files and folders. There were many different ways to manage your files and folders in Ionic 1x. Usually the default structure you will see in the Ionic 1x build can be used in many small projects where we don't have to deal with many files and for small projects; mainly two structures are followed for Ionic 1x based projects: **byFolderType** and **byFeatureType**. With Ionic 3, one of the best things is that we have the project structure set up in a modular way where you will have pages, providers, pipes, theme folders, and respective subsequent folders in them. Let's go through some important files inside Ionic 3 projects:

`./src`

In this folder we have our entire application code base and it is the most important:

`./node_modules`

Here we have all our dependencies and NPM packages that are required to run the application:

`./platforms`

Here we have specific platform-based folder entries, which our app is using. When we run the `ionic platform add` command, you will see a specific folder created inside the `platform` folder:

`./plugins`

All the plugins used in our application are hosted here:

`./resources`

An entire set of icons and splash screens are inside this folder, organized according to various platforms and devices:

`./www`

Here is our `index.html` file, and after compilation of the code base in the `src` folder bundle, files, images, and SASS compiled css files are placed inside `www`:

`./www/index.html`

This is the entry point to our application or we can say every hybrid application. We mainly have scripts and style sheets declared in this file to run the application. As in Ionic 1 we use `ng-app`, here in Ionic 3 we check for `<ion-app></ion-app>` inside your `index.html` file:

`./package.json`

This defines the project metadata and dependencies that will be added to the `node_modules` folder. It also has information about the Cordova platform and plugins used in the application:

`./ionic.config.json`

This file contains project-specific settings such as names, IDs, proxies, and other information:

```
./src/app/main.ts
```

This is where we start towards the code base and also we bootstrap our application inside this file:

```
./src/app/app.module.ts
```

In this file we declare the pages, directives, and providers used inside our application:

```
./src/app/app.component.ts
```

We set the root page here and root-component is called first, which controls the application, similar to what the app.js file does in Ionic 1. However, this component is not different from other components in our application; it is just declared in the app folder. As we start to build applications in upcoming chapters, we will get to know more closely how we work around all these files and code bases.

Theming up SASS

Ionic has support for SASS (CSS extension language) to create, customize, and maintain CSS. It eases the process of customizing the existing colors and themes of Ionic for specific platforms. Ionic 3 as a default has SASS setup and you will find that inside the theme folder there will be an variable.scss file. In this file we will be doing custom color changes and overriding platform variables.
Earlier in Ionic 1x, the application was hooked to the Ionic's precompiled files, which you can find at the www/lib/ionic/css directory, and file resources and paths are linked in index.html. Previously, in Ionic 1x, we used to set up SASS using CLI:

```
$ ionic setup sass
```

This used to automatically remove other file paths and uncomment the ionic.app.css files used for SASS styling inside index.html. Now with Ionic 3 we don't have to set up SASS, it comes by default when we start or create an application.

Customizing

Now the basic setup is ready for SASS and also while running `ionic serve` you will add all custom styling and SASS related changes in all different files for each of the components or pages. Let's start with changing one of the color variables by default present in the `variable.scss` file:

```
    /*
To customize the look and feel of
    Ionic, you can override the variables
For example, you might change some of
    the default colors:

$colors: (
  primary:    #387ef5,
  secondary:  #32db64,
  danger:     #f53d3d,
  light:      #f4f4f4,
  dark:       #222,
  favorite:   #69BB7B,
  twitter:    #53ACEB,
  github:     #000000,
  instagram:  #235D8D
);

// Fastest way to change the theme of
        your Ionic app is to set new value
        for primary color
```

You can see that we have added custom colors such as Twitter and GitHub. We can further customize it by supplying base and contrast properties:

```
    $colors: (
facebook:(
  base: #3b5998,
  contrast: #ffffff
  )
)
```

You can now use this color key as a property to many components, such as `buttons`:

```
<button facebook>Share on Facebook</button>
```

SASS is really powerful and it speeds up the CSS development process. It acts as a Swiss army knife for CSS, where we can do multiple things with minimum lines of code and then with just some changes we can easily change the entire look and feel of the application. There are so many examples of how we can use SASS, but as we are more into Ionic development, I recommend going through some good SASS documentation and Ionic 3 theming docs once so that detailed information will be provided.

Automatically creating icons and splash screen resources

Ionic tools are so beautiful that now you don't have to struggle like we all used to, some years back in hybrid development. You can avoid the headache of adding icons, splash screens, and so on for every platform with different sizes. Now we don't have to deal with all this stress and Ionic have made it possible for us with a single command:

```
$ ionic resources
```

Ionic automatically crops, resizes, and creates icons and splash screens from source images for each platform and it does this in different sizes for different devices such as mobiles, tablets, and so on. All these are generated on Ionic's resizing and cropping servers without any overload from installing libraries or plugins on local machines.

Image sizes and specifications

Ionic resources have some specification of the source image of the icon and splash screen. Images can be either `.png` files, `Photoshop.psd` files, or Illustrator `.ai` files. There is a minimum size requirement for the source images for both. An icon's minimum size should be 192x192 px, and it should not have rounded corners. In case of splash screen, the minimum requirement is 2208x2208 px so that for every platform Ionic can prepare resources. The splash screen's artwork should roughly fit within a center square (1200x1200 px). Additionally, when the orientation preference configuration is set to either landscape or portrait mode, then only the necessary images will be generated.

For creating just icons or splash screen, the `ionic resources` command has two flags that allow you to create just icons or splash screens, not both:

```
$ ionic resources --icon
$ ionic resources --splash
```

If a proper size is not provided for the source files, this will only create resources that are less than the size of the source image. For example, there might be chances that resources for tablets or high resolution screens are not generated.

Platform specific resources

Ionic provides support for building icons and splash screen resources for various platforms and devices. For building resources, we just need to place the source image inside the `resources` folder with the name `icon.png` and `splash.png`. This way you will get the extracted icons and splash screens for each platform, such as a Native rounded corners icon for iOS, and transparent background for an Android icon.

To summarize the steps:

1. Add files to the `resources` folder naming `icon.png` and `splash.png`. (`.psd` and `.ai` can also be used).
2. Make sure the minimum size requirements for icon are 192x192 px and for splash 2208x2208 px.
3. Run the `ionic resources` command from the CLI.

 Different platforms have different ways of styling icons, for example, iOS will apply its rounded corners, that is why we recommend the `Ionic` source file to be without rounded corners. Also, using the `ionic resources` command will require at least Cordova 3.6 or more.

Adding Crosswalk browser

Older Android versions (4.0 - 4.3) that stock web view have low performance and lack many of the latest HTML5, CSS3, and JS features. You will see a lot of difference when you deploy your application on the latest Android 7.0 and in older versions. Here is when Crosswalk comes into the frame; Crosswalk gives the latest web view aligned with Chrome on Android. This increases the performance of both HTML/CSS rendering and JavaScript performance ten times. It reduces fluctuations and fragmentation among devices. Another set of features that Crosswalk brings is access to webGL, WebRTC, CSS3 features, and various APIs. It provides improved performance and debugging applications become really easy. The Cordova Crosswalk plugin helps you easily add the Crosswalk browser in your application:

```
$ ionic cordova:plugin add cordova-plugin-crosswalk-
```

```
webview
```

Currently, supported browsers are Crosswalk and Crosswalk-lite for Android. You can use Crosswalk lite mode by passing a variable flag:

```
$ ionic cordova:plugin add cordova-plugin-crosswalk-
  webview --variable XWALK_MODE="lite"
```

Please take care that running these step will replace the default browser. Although you can anytime revert back by uninstalling the plugin and build again. The following are some advantages of using Crosswalk:

- Gain in performance
- Reduced fluctuations and fragmentation
- Ease of debugging
- HTML5 and CSS3 features
- Access to webRTC, webGL, web Sockets, and so on

Another thing to note is that after you have added the Crosswalk browser you will see the size of your APK increase by around 15-20 MB and increased of size on disk when installed around 50 MB.

Ionic CLI tasks

Ionic is improving day by day and many new features are coming to the platform.

Local development with Ionic serve

During the development process we need continuous testing. With Hybrid development you have the advantage that you can test your application in multiple ways:

- In desktop browsers
- iOS/Android simulators
- Mobile browsers
- Installing as a Native application

So for development purposes, testing in a desktop browser is the quickest and easiest. Many moving parts can be tested on desktop browsers. For running on desktop browsers, the Ionic CLI provides a command for it:

```
    $ ionic serve [options]
$ ionic serve --lab
```

Adding platforms

The `ionic platform` command adds and removes platforms:

```
$ ionic cordova:platform <action> <PLATFORM>
[options]
$ ionic cordova:platform add android
$ ionic cordova:platform remove ios
  Running applications on devices
```

Running application on devices

The `ionic run` command helps in running the application with the connected device:

```
$ ionic cordova:run [options] <PLATFORM>
$ ionic cordova:run android
  There is also an option of Emulate. Emulate an
  Ionic
  project on a simulator or emulator.
$ ionic cordova:emulate [options] <PLATFORM>
```

The list of available tasks has been increasing in Ionic CLI. The best way to get in touch with all the latest tasks with extended details can be found by using `ionic --help`:

```
    $ ionic--help
```

This will list all the tasks with options and flags for each. You can easily try them all on a test project and this will help you understanding them more closely.

Uploading and sharing Ionic application

While developing the application you now have features to share your application with a testing team or friends for reviewing. The Ionic platform provides this feature and the best part is that you don't even need to build locally and then send everyone APK files. Uploading and sharing an application is not just some easy steps. Ionic for organization is another service available for collaboration between employees of an organization.

Ionic upload

Ionic upload is used for uploading new snapshot of your app to your Ionic account. Before this command you need to run `ionic link` which will add `app_id` and create an application on Ionic Cloud:

```
$ ionic upload
```

Arguments or flags for the upload command:

- `[--deploy]`: Deploy snapshot to specific channel
- `[--note]`: The note to signify the upload description

Generating components

The ionic generate command create new pipes, components, pages, directive and providers. Also, this will create entry inside you app.module.ts for example if you create have created a provider it will be imported in `app.module.ts` and injected in providers. Once you create any of the page, component, directive, or pipe you will see a folder `create` with it TypeScript file where you can now start writing your logic:

```
$ ionic generate [type] [name]
```

- `[--type]`: Type of generator such as page, directive, and pipe
- `[--name]`: Name of the component generated

```
// Some example commands
$ ionic generate pipe NumberPipe
$ ionic generate page About
```

Ionic share

While you are building your application, you can directly test it on the Ionic View application where you can mention the app ID and it will automatically download your application from the `ionic.io` server where you uploaded your application:

```
$ ionic share <email>
```

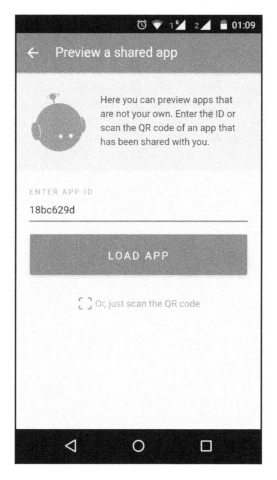

After you are done with uploading and sharing your application with users with specific e-mail addresses, the person you sent the invitation to will get an e-mail with a link to view the app. You can also test your application directly on the Ionic View application, which is available on Google Play Store and Apple Store. You can also directly share your app ID to enter building your application, so you can directly test it on the Ionic View application. The preceding screenshot shows how you can add an app ID and test the application.

Summary

We've now covered the get started part of Ionic and have gone through various aspects of what to take care of while building a large application. We now know how important it is to organize your files and folders in a project. We also discussed various setups, such as customizing applications through SASS, automatically creating icons and splash screens for various platforms, and uploading/sharing applications. Later in the chapter we went on to cover many Ionic CLI commands that help us quickly develop and debug applications. Ionic CLI is advancing day by day and many new features and tasks are coming. What we all should do is regularly update Ionic CLI and check for new features so that in our next application we can take advantage of them. We are now prepared to dig more deeply into Ionic and its components. In `Chapter 2`, *Ionic Components*, we will be going around with the Ionic components and this is where we will move onto real application building where we will be using lists, virtual scrolling, localization, navigation, and many other components. This will help us make complex applications where we will have many tasks at hand. UI customization is another aspect we will look into, as almost every application will have a different design and we can easily build them over the current Ionic components with small tweak. Another important aspect we will look into is the custom modules built by the Ionic community, which we can directly use inside our application.

2
Ionic Components

In this chapter, we will be going through various Ionic components and APIs that are necessary while building large applications. Ionic provides various built-in JavaScript components that we can utilize intelligently. There are many APIs provided by Ionic that we are not really aware of and here we will learn how we can efficiently use these in our application to increase the performance of the application and make it production ready. We will also be starting here with an e-commerce application named vPlanet Commerce; I named this as one of my product ideas where we will be taking e-commerce stores to mobile. We will add progressive web app support to the application later in this book so that we have a complete package. To make this chapter interesting, I will integrate different Ionic components and APIs according to what an e-commerce app requires while building. This way you will have an entire e-commerce application ready.

The topics that we will be covering in this chapter are:

- Building vPlanet Commerce
- Ionic 3 components
- Ionic APIs and custom modules
- Lazy Loading with Ionic 3

Building vPlanet Commerce

We have been talking for a while about building an e-commerce application; now we will be getting started with that. As previously mentioned, it is named vPlanet Commerce. The vPlanet Commerce app is an e-commerce app that will demonstrate various Ionic components integrated inside the application and also some third-party components built by the community.

As we move forward to the Chapter 3, *Ionic Native and Plugins*, we will be integrating Ionic Cloud services such as auth, Ionic DB, deploy, and so on. This will help us complete the application and make it look like a professional application. Let's start by creating the application from scratch using the sidemenu template:

```
gaurav@gaurav-thinkpad:~/projects$ ionic start vplanet-commerce sidemenu
✓ Creating directory /home/gaurav/projects/vplanet-commerce - done!
✓ Downloading 'sidemenu' starter template - done!
✓ Updating project dependencies to add required plugins - done!
✓ Creating configuration file for the new project - done!
✓ Executing: npm install within the newly created project directory - done!

♫ ♪ ♫ ♪  Your Ionic app is ready to go!  ♫ ♪ ♫ ♪

Run your app in the browser (great for initial development):
  ionic serve

Run on a device or simulator:
  ionic cordova:run ios

Test and share your app on a device with the Ionic View app:
  http://view.ionic.io

? Link this app to your Ionic Dashboard to use tools like Ionic View? Yes
[OK]
      Run ionic link to link to the app.

Go to your newly created project: cd /home/gaurav/projects/vplanet-commerce

gaurav@gaurav-thinkpad:~/projects$
```

You now have the basic application ready based on the side menu template. The next step I took is to take reference from ionic-conference-app for building the initial components of the application, same as the walkthrough.

Let's create a walkthrough component via CLI to generate a command:

```
$ ionic g page walkthrough
```

As we get started with the walkthrough component, we need to add logic to show the walkthrough component only the first time when the user installs the application:

```
// src/app/app.component.ts
// Check if the user has already seen the walkthrough
    this.storage.get('hasSeenWalkThrough').then((hasSeenWalkTh rough) => {
        if (hasSeenWalkThrough) {
          this.rootPage = HomePage;
        } else {
          this.rootPage = WalkThroughPage;
        }
        this.platformReady();
    })
```

So, we store a `boolean` value while checking if the user has seen the walkthrough for the first time or not. Another important thing we did was create `Events` for login and logout, so that when the user logs into the application we can update `Menu` items accordingly or any other data manipulation to be done:

```
// src/app/app.component.ts

export interface PageInterface {
  title: string;
  component: any;
  icon: string;
  logsOut?: boolean;
  index?: number;
  tabComponent?: any;
}
export class vPlanetApp {
    loggedInPages: PageInterface[] = [
      { title: 'account', component: AccountPage, icon:
        'person' },
      { title: 'logout', component: HomePage, icon: 'log-
```

```
        out', logsOut: true }
    ];
  loggedOutPages: PageInterface[] = [
    { title: 'login', component: LoginPage, icon: 'log-in'
    },
    { title: 'signup', component: SignupPage, icon:
      'person-add' }
    ];

    listenToLoginEvents() {
        this.events.subscribe('user:login', () => {
          this.enableMenu(true);
        });

        this.events.subscribe('user:logout', () => {
          this.enableMenu(false);
        });
    }

    enableMenu(loggedIn: boolean) {
        this.menu.enable(loggedIn, 'loggedInMenu');
        this.menu.enable(!loggedIn, 'loggedOutMenu');
    }

  // For changing color of Active Menu

    isActive(page: PageInterface) {
        if (this.nav.getActive() &&
     this.nav.getActive().component === page.component) {
          return 'primary';
        }
        return;
    }

  }
```

Next we have multiple `app.template.html`,and we have multiple `<ion-menu>` items depending upon whether user is loggedin or logout:

```
// src/app/app.template.html
<ion-split-pane>
<!-- logged out menu -->
<ion-menu id="loggedOutMenu" [content]="content">

  <ion-header>
  <ion-toolbar>
  <ion-title>{{'menu' | translate}}</ion-title>
  </ion-toolbar>
```

```
</ion-header>

<ion-content class="outer-content">

<ion-list>
<ion-list-header>
{{'navigate' | translate}}
</ion-list-header>
<button ion-item menuClose *ngFor="let p of appPages"
 (click)="openPage(p)">
<ion-icon item-left [name]="p.icon" [col-
 or]="isActive(p)"></ion-icon>
{{ p.title | translate }}
</button>
</ion-list>

<ion-list>
<ion-list-header>
{{'account' | translate}}
</ion-list-header>
<button ion-item menuClose *ngFor="let p of
 loggedOutPages" (click)="openPage(p)">
<ion-icon item-left [name]="p.icon" [col-
 or]="isActive(p)"></ion-icon>
{{ p.title | translate }}
</button>
<button ion-item menuClose *ngFor="let p of
 otherPages" (click)="openPage(p)">
<ion-icon item-left [name]="p.icon" [col-
 or]="isActive(p)"></ion-icon>
{{ p.title | translate }}
</button>
</ion-list>
</ion-content>

</ion-menu>

<!-- logged in menu -->
<ion-menu id="loggedInMenu" [content]="content">

<ion-header>
<ion-toolbar>
<ion-title>Menu</ion-title>
</ion-toolbar>
</ion-header>

<ion-content class="outer-content">
```

```
<ion-list>
<ion-list-header>
{{'navigate' | translate}}
</ion-list-header>
<button ion-item menuClose *ngFor="let p of appPages"
  (click)="openPage(p)">
<ion-icon item-left [name]="p.icon" [col-
  or]="isActive(p)"></ion-icon>
{{ p.title | translate }}
</button>
</ion-list>

<ion-list>
<ion-list-header>
{{'account' | translate}}
</ion-list-header>
<button ion-item menuClose *ngFor="let p of
  loggedInPag-es" (click)="openPage(p)">
<ion-icon item-left [name]="p.icon" [col-
  or]="isActive(p)"></ion-icon>
{{ p.title | translate }}
</button>
<button ion-item menuClose *ngFor="let p of
  otherPages" (click)="openPage(p)">
<ion-icon item-left [name]="p.icon" [col-
  or]="isActive(p)"></ion-icon>
{{ p.title | translate }}
</button>
</ion-list>

</ion-content>

</ion-menu>
```

As our app starts mainly from `app.template.html` so we declare `rootPage` here. Recently, `SplitPane` component was released which helps creating multi-view layout, which is really helpful running the same application on different devices and screen sizes:

```
<!-- main navigation -->
<ion-nav [root]="rootPage" #content
  swipeBackEnabled="false" main></ion-nav>
</ion-split-pane>
```

Let's now look into what pages, services, and filters we will be having inside our app. Rather than mentioning it as a bullet list, the best way to know this is going through the `app.module.ts` file, which has all the declarations, imports, `entrycomponents`, and `providers`:

```
// src/app/app.modules.ts

import { BrowserModule } from '@angular/platform-browser';
import { NgModule, ErrorHandler } from '@angular/core';
import { IonicApp, IonicModule, IonicErrorHandler, DeepLinkCon-fig
} from 'ionic-angular';
import { TranslateModule, TranslateLoader,
        TranslateStaticLoad-er } from 'ng2-
        translate/ng2-translate';
import { HttpModule, Http } from '@angular/http';
import { CloudSettings, CloudModule } from '@ionic/cloud-angular';
import { Storage } from '@ionic/storage';
import { vPlanetApp } from './app.component';


import { AboutPage } from '../pages/about/about';
import { PopoverPage } from '../pages/popover/popover';
import { AccountPage } from '../pages/account/account';
import { LoginPage } from '../pages/login/login';
import { SignupPage } from '../pages/signup/signup';
import { WalkThroughPage } from '../pages/walkthrough/walkthrough';
import { HomePage } from '../pages/home/home';
import { CategoriesPage } from
        '../pages/categories/categories';
import { ProductsPage } from
        '../pages/products/products';
import { ProductDetailPage } from '../pages/product-
        detail/product-detail';
import { WishlistPage } from
        '../pages/wishlist/wishlist';
import { ShowcartPage } from
        '../pages/showcart/showcart';
import { CheckoutPage } from
        '../pages/checkout/checkout';
import { ProductsFilterPage } from '../pages/products-
        filter/products-filter';
import { SupportPage } from '../pages/support/support';
import { SettingsPage } from
        '../pages/settings/settings';
import { SearchPage } from '../pages/search/search';
import { UserService } from '../providers/user-
        service';
import { DataService } from '../providers/data-
        service';
import { OrdinalPipe } from '../filters/ordinal';

// 3rd party modules
import { Ionic2RatingModule } from 'ionic2-rating';
```

```
// import ionic native packages
import { SplashScreen } from '@ionic-native/splash-
        screen';
import { Deeplinks } from '@ionic-native/deeplinks';
import { TextToSpeech } from '@ionic-native/text-to-
        speech';
import { AppRate } from '@ionic-native/app-rate';
import { GoogleAnalytics } from '@ionic-native/google-
        analytics';
import { Camera } from '@ionic-native/camera';
import { CallNumber } from '@ionic-native/call-number';
import { SocialSharing } from '@ionic-native/social-
        sharing';
import { Push } from '@ionic-native/push';

export function createTranslateLoader(http: Http) {
 return new TranslateStaticLoader(http, './assets/i18n', '.json');
}

// Configure database priority
export function provideStorage() {
 return new Storage(['sqlite', 'indexeddb', 'localstorage'], {
name: 'vplanet' })
}

const cloudSettings: CloudSettings = {
 'core': {
 'app_id': 'f8fec798'
 }
};

// Deeplink Configuration
export const deepLinkConfig: DeepLinkConfig = {
 links: [
 { component: HomePage, name: 'Home Page', segment: ''
 },
 { component: CategoriesPage, name: 'Categories Page',
   seg-ment: 'categories' },
 { component: ProductsPage, name: 'Categories Product
   Page', segment: 'categories/:categoryId' },
 { component: ProductDetailPage, name: 'Product Details
   Page', segment: 'products/:productId' },
 { component: WishlistPage, name: 'Wishlist Page',
   segment: 'wishlist' },
 { component: ShowcartPage, name: 'Showcart Page',
   segment: 'cart' },
 { component: SupportPage, name: 'Support Page',
   segment: 'feedback' },
```

```
    { component: SettingsPage, name: 'About Page',
      segment: 'settings' },
    { component: AboutPage, name: 'About Page', segment:
      'about' },
    { component: LoginPage, name: 'Login Page', segment:
      'login' },
    { component: SignupPage, name: 'Signup Page', segment:
      'signup' },
    { component: AccountPage, name: 'Account Page',
      segment: 'account' }
    ]
};

@NgModule({
  declarations: [
  vPlanetApp,
  AboutPage,
  AccountPage,
  LoginPage,
  PopoverPage,
  SignupPage,
  WalkThroughPage,
  HomePage,
  CategoriesPage,
  ProductsPage,
  ProductsFilterPage,
  ProductDetailPage,
  SearchPage,
  WishlistPage,
  ShowcartPage,
  CheckoutPage,
  SettingsPage,
  SupportPage,
  OrdinalPipe,
  ],
  imports: [
  BrowserModule,
  HttpModule,
  IonicModule.forRoot(vPlanetApp, {locationStrategy: 'hash'},
deepLinkConfig),
  Ionic2RatingModule,
  TranslateModule.forRoot({
  provide: TranslateLoader,
  useFactory: createTranslateLoader,
  deps: [Http]
  }),
  IonicStorageModule.forRoot({ useFactory:
  provideStorage }),
```

```
CloudModule.forRoot(cloudSettings)
],
bootstrap: [IonicApp],
entryComponents: [
vPlanetApp,
AboutPage,
AccountPage,
LoginPage,
PopoverPage,
SignupPage,
WalkThroughPage,
HomePage,
CategoriesPage,
ProductsPage,
ProductsFilterPage,
ProductDetailPage,
SearchPage,
WishlistPage,
ShowcartPage,
CheckoutPage,
SettingsPage,
SupportPage
],
providers: [
SplashScreen,
Deeplinks,
TextToSpeech,
AppRate,
GoogleAnalytics,
Camera,
CallNumber,
SocialSharing,
Push,
{provide: ErrorHandler, useClass: IonicErrorHandler},
{ provide: Storage, useFactory: provideStorage },
UserService,
DataService
]
})

export class AppModule {}
```

Furthermore, I would suggest you look at the code in this chapter for the vPlanet Commerce app, which will help you get a good idea of what pages and components we will be using in our application. As this chapter mainly targets various Ionic components, we will be using a lot of these components in our application.

Ionic components

There are many Ionic JavaScript components that we can effectively use while building our application. What's best is to look around for features that we will need in our application. Let's get started with the home page of our e-commerce application, which will have an image slider with banners on it.

Slides

The slides component is a multi-section container that can be used in multiple scenarios, such as a tutorial view or banner slider. The `<ion-slides>` component has multiple `<ion-slide>` elements, which can be dragged or swiped left/right. Slides have multiple configuration options available, which can be passed in the `ion-slides` such as autoplay, pager, direction: vertical/horizontal, initial slide, and speed.

Using slides is really simple as we just have to include them inside our `home.html`; no dependency is required for this to be included in the `home.ts` file:

```
// Banner Slider in src/pages/home/home.html

<ion-slides pager #adSlider (ionSlideWillDid-
 Change)="logLenthtrackEvent()" style="height: 250px">
 <ion-slide *ngFor="let banner of banners">
 <img [src]="banner">
 </ion-slide>
</ion-slides>

// Defining banners image path

export class HomePage {

 products: any;
 banners: String[];

 constructor() {
 this.banners = [
 'assets/img/banner-1.webp',
 'assets/img/banner-2.webp',
 'assets/img/banner-3.webp'
 ]
 }
}
```

As you can see from the preceding code, there are multiple properties available for ion-slides, such as pager, autoplay, direction, and so on:

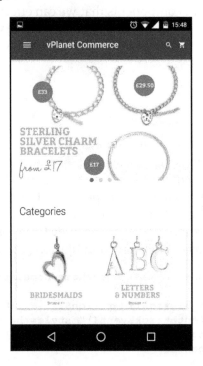

If you look at the preceding screenshot, you can see we have used white background banners. Actually, just as a suggestion, try to use the same background color in banners that are inside the application. This way the banners will come up nice in low or high resolution mobiles and tabs:

```
// src/pages/home/home.ts

import { Component, ViewChild } from '@angular/core';

import { Slides } from 'ionic-angular';

class HomePage {
 @ViewChild('adSlider') slider: Slides;

 logLenth() {
 console.log(this.slider.length());
 }
 trackEvent() {
 let active = this.slider.getActiveIndex();
 this.platform.ready().then(() => {
```

```
// Google Analytics Plugin, will introduce about this
   in next chapter
this.ga.trackEvent("Slider", "Slider-Changed", "Label", active);
});
}


}
```

There are multiple methods available such as `slideTo`, `slideNext`, `getActiveIndex`, and so on. You can find all these on Ionic documentation at `https://ionicframework.com/docs/v2/api/components/slides/Slides/`.

Lists

Lists are one of the most used components in many applications. Inside lists we can display rows of information. We will be using lists multiple times inside our application, such as on the `categories` page where we are showing multiple subcategories:

```
// src/pages/categories/categories.html

<ion-content class="categories">

  <ion-list-header *ngIf="!categoryList">Fetching
  Categories ....</ion-list-header>

  <ion-list *ngFor="let cat of categoryList">

   <ion-list-header>{{cat.name}}</ion-list-header>

   <ion-item *ngFor="let subCat of cat.child">
     <ion-avatar item-left>
       <img [src]="subCat.image">
     </ion-avatar>
     <h2>{{subCat.name}}</h2>
     <p>{{subCat.description}}</p>
     <button ion-button clear item-right
       (click)="goToProducts(subCat.id)">View</button>
   </ion-item>

  </ion-list>

</ion-content>
```

We are using an avatar list here with a header on top of each category. There are multiple types of lists available such as multi-line, thumbnail, and sliding lists also. We will be integrating the sliding list on the showcart page further in this chapter:

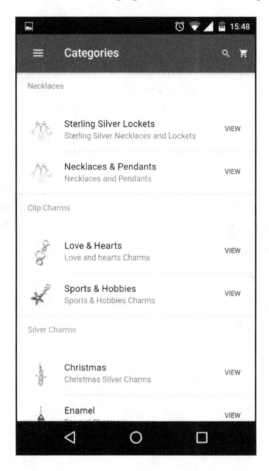

As you can see, the preceding screenshot shows what the avatar list will look like on the categories page. We should make sure that, before we plan to use lists in our application, we have a look at the documentation, as this helps us visualize which type of list can be utilized best in various different applications.

Loading and toast

The loading component can be used to indicate some activity while blocking any user interactions. One of the most common cases of using the Loading component is HTTP/ calls to the server, as we know it takes time to fetch data from the server, until then for good user experience we can show some content showing Loading .., or Login wait .., for login pages.

Toast is a small popup that provides feedback, usually used when some action is performed by the user. Ionic 3 now provides the toast component as part of its library; previously we had to use the native Cordova plugin for toasts, which can also be used now too.

Loading and toast components both have a method Create. We have to provide options while creating these components:

```
// src/pages/login/login.ts

import { Component } from '@angular/core';
import { NgForm } from '@angular/forms';
import { NavController, LoadingController, ToastController, Events } from
'ionic-angular';
import { SignupPage } from '../signup/signup';
import { HomePage } from '../home/home';

import { Auth, IDetailedError } from '@ionic/cloud-angular';

import { UserService } from '../../providers/user-service';

@Component({
  selector: 'page-user',
  templateUrl: 'login.html'
})
export class LoginPage {
  login: {email?: string, password?: string} = {};

  submitted = false;

  constructor(public navCtrl: NavController,
    public loadingCtrl: LoadingController,
    public auth: Auth,
    public userService: UserService,
    public toastCtrl: ToastController,
    public events: Events) { }

  onLogin(form: NgForm) {
    this.submitted = true;
```

```
        if (form.valid) {
          // start Loader
          let loading = this.loadingCtrl.create({
            content: "Login wait...",
            duration: 20
          });
          loading.present();

          this.auth.login('basic', this.login).then((result)
           => {
            // user is now registered
            this.navCtrl.setRoot(HomePage);
            this.events.publish('user:login');
            loading.dismiss();
            this.showToast(undefined);
          }, (err: IDetailedError<string[]>) => {
            console.log(err);
            loading.dismiss();
            this.showToast(err)
          });
        }
      }

  showToast(response_message:any) {
      let toast = this.toastCtrl.create({
        message: (response_message ? response_message : "Log In
Successfully"),
        duration: 1500
      });
      toast.present();
  }

  onSignup() {
      this.navCtrl.push(SignupPage);
  }
}
```

As you can see from the preceding code, creating a loader and toast is almost similar at code level. The options provided while creating are also similar; we have used loader here while logging in and toast after that to show the desired message.

Setting the duration option is good, because if the case loader is dismissed or not handled properly in code then we will block the user from any further interactions on the app. In HTTP calls to the server we might get connection issues or failure cases; in that scenario it may end up blocking users.

Tabs versus segments

Tabs are the easiest way to switch between views and organize content at a higher level. On the other hand, a segment is a group of buttons and it can be treated as a local switch `Tab` inside a particular component mainly used as a filter. With `Tabs` we can build, quick access bar in the footer where we can place menu options such as home, favorites, and cart. This way we can have one click access to these pages or components. On the other hand, we can use segments inside the `account` component and divide the data displayed in three segments, `profile`, `orders`, and `wallet`:

```
// src/pages/account/account.html

<ion-header>
  <ion-navbar>
    <button menuToggle>
      <ion-icon name="menu"></ion-icon>
    </button>
    <ion-title>Account</ion-title>
  </ion-navbar>
  <ion-toolbar [color]="isAndroid ? 'primary' : 'light'"
   no-border-top>
    <ion-segment [(ngModel)]="account" [color]="isAndroid
     ? 'light' : 'primary'">
      <ion-segment-button value="profile">
        Profile
      </ion-segment-button>
      <ion-segment-button value="orders">
        Orders
      </ion-segment-button>
      <ion-segment-button value="wallet">
        Wallet
      </ion-segment-button>
    </ion-segment>
  </ion-toolbar>
</ion-header>

<ion-content class="outer-content">
  <div [ngSwitch]="account">

    <div padding-top text-center *ngSwitchCase="'profile'"
    >
      <img src="http://www.gravatar.com/avatar?
      d=mm&s=140">
      <h2>{{username}}</h2>

      <ion-list inset>
        <button ion-item (click)="updatePicture()">Update
```

```
        Picture</button>
      <button ion-item (click)="changePassword()">Change
      Password</button>
      <button ion-item
       (click)="logout()">Logout</button>
    </ion-list>

  </div>

  <div padding-top text-center *ngSwitchCase="'orders'"
  >
    // Order List data to be shown here
  </div>

  <div padding-top text-center *ngSwitchCase="'wallet'">
    // Wallet statement and transaction here.
  </div>

  </div>
</ion-content>
```

This is how we define a segment in Ionic; we don't need to define anything inside the `TypeScript` file for this component. On the other hand, with tabs we have to assign a component for each tab and we can also access its methods via the `Tab` instance. Just to mention, we haven't used Tabs inside our e-commerce application, as we are using the side menu. One good example will be to look at the ionic-conference-app at `https://github.com/driftyco/ionic-conference-app`. You will find both `sidemenu` and `Tabs` in the single application:

```
/
// We currently don't have Tabs component inside our e-commerce application
// following is the sample code about how we can integrate it.

<ion-tabs #showTabs tabsPlacement="top" tabsLayout="icon-top"
color="primary">
  <ion-tab [root]="Home"></ion-tab>
  <ion-tab [root]="Wishlist"></ion-tab>
  <ion-tab [root]="Cart"></ion-tab>
</ion-tabs>

import { HomePage } from '../pages/home/home';
import { WishlistPage } from '../pages/wishlist/wishlist';
import { ShowcartPage } from '../pages/showcart/showcart';

export class TabsPage {

  @ViewChild('showTabs') tabRef: Tabs;

  // this tells the tabs component which Pages
  // should be each tab's root Page

  Home = HomePage;
  Wishlist = WishlistPage;
  Cart = ShowcartPage;

  constructor() {

  }

  // We can access multiple methods via Tabs instance
  // select(TabOrIndex), previousTab(trimHistory),
     getByIndex(index)
  // Here we will console the currently selected Tab.

  ionViewDidEnter() {
    console.log(this.tabRef.getSelected());
  }
}
```

Properties can be checked in the documentation at
(`https://ionicframework.com/docs/v2/api/components/tabs/Tabs/`), as there are many
properties available for `Tabs`, such as mode, color, `tabsPlacement`, and `tabsLayout`.
Similarly, we can configure some tabs properties at the config level also; you will find what
properties you can configure globally or for a specific platform at
`https://ionicframework.com/docs/v2/api/config/Config/`.

Alerts

Alerts are the components provided in Ionic for showing trigger alerts, confirm, prompts, or
some specific actions. AlertController can be imported from ionic-angular, which allows us
to programmatically create and show alerts inside the application. One thing to note here is
these are JavaScript popups and not the native platform popups. There is a Cordova plugin
called `cordova-plugin-dialogs`
(`https://ionicframework.com/docs/v2/native/dialogs/`), which you can use if native
dialog UI elements are required.

Currently there are five types of alerts that we can show in the Ionic app, basic alert, prompt
alert, confirmation alert, radio, and checkbox alerts.

A radio alert `inside src/pages/products/products.html` is for sorting products:

```
<ion-buttons>
  <button ion-button full clear (click)="sortBy()">
    <ion-icon name="menu"></ion-icon>Sort
  </button>
</ion-buttons>

// onClick we call sortBy method
// src/pages/products/products.ts
import { NavController, PopoverController, ModalController, AlertController
} from 'ionic-angular';

export class ProductsPage {

  constructor(
       public alertCtrl: AlertController
  ) {

    sortBy() {
        let alert = this.alertCtrl.create();
        alert.setTitle('Sort Options');

        alert.addInput({
          type: 'radio',
          label: 'Relevance',
```

```
      value: 'relevance',
      checked: true
    });
    alert.addInput({
      type: 'radio',
      label: 'Popularity',
      value: 'popular'
    });
    alert.addInput({
      type: 'radio',
      label: 'Low to High',
      value: 'lth'
    });
    alert.addInput({
      type: 'radio',
      label: 'High to Low',
      value: 'htl'
    });
    alert.addInput({
      type: 'radio',
      label: 'Newest First',
      value: 'newest'
    });

    alert.addButton('Cancel');
    alert.addButton({
      text: 'OK',
      handler: data => {
          console.log(data);
       // Here we can call server APIs with sorted data
       // using the data which user applied.
        }
    });

    alert.present().then(() => {
      // Here we place any function that
   // need to be called as the alert in opened.
    });
  }

}
```

Here are the **Cancel** and **OK** buttons. We have used this here for sorting the products according to relevance, price, or other sorting values.

We can prepare custom alerts also, where we can mention multiple options. As in the preceding example, we have five radio options; similarly we can even add a text input box for taking some inputs and submit it. Other than this, while creating alerts remember that there are alerts, input, and button options properties for all the alerts present in the `AlertController` component (https://ionicframework.com/docs/v2/api/components/alert/AlertController/).

The following are some alert options:

- `title: // string`: Title of the alert
- `subTitle: // string (optional)`: Sub-title of the popup
- `Message: // string`: Message for the alert
- `cssClass: // string`: Custom CSS class name
- `inputs: // array`: Set of inputs for the alert
- `buttons // array (optional)`: Array of buttons

Cards and badges

Cards are one of the important components used more often in mobile and web applications. The reason why cards are so popular is because it's a great way to organize information and also get the users' access to a large quantity of information on smaller screens also. Cards are really flexible and responsive due to all these reasons they are adopted very quickly by developers and companies. We will also be using cards inside our application on the home page itself for showing popular products. Let's see what different types of cards Ionic provides in its library:

- Basic cards
- Cards with headers and footers
- Cards lists
- Cards images
- Background cards
- Social and map cards
- Social and Map cards are advanced cards that are built with custom CSS

We can develop similar advanced cards also:

```
// src/pages/home/home.html

<ion-card>
  <img [src]="prdt.imageUrl"/>
  <ion-card-content>
    <ion-card-title no-padding>
  {{prdt.productName}}
  </ion-card-title>
    <ion-row no-padding class="center">
     <ion-col>
       <b>{{prdt.price | currency }}   </b><span
class="discount">{{prdt.listPrice | currency}}</span>
     </ion-col>
    </ion-row>
  </ion-card-content>
</ion-card>
```

We have used an image card here with an image on top, and next we have **favorite** and **view** button icons. Similarly, we can use different types of cards where ever it's required. Also, at the same time we can customize our cards and mix two types of cards using their specific CSS classes or elements.

Badges are small components used to show small information, for example, showing the number of items in the **cart** above the **cart** icon. We have used these in our e-commerce application for showing the ratings of products:

```
<ion-badge width="25">4.1</ion-badge>
```

Ionic APIs and custom modules

Until now, we have worked on multiple Ionic components. To support these components we have Component APIs for classes such as checkbox, toggle, or item, which show how to use them. We have already used many APIs while we used the components. Now we will look further at all the useful APIs we have that can be used inside our e-commerce application. Also, another important aspect is using custom modules, which will further enhance the application by adding localization, making custom pipes, and so on.

Complex grids

We will be starting with one of the common problems many developers face while building applications in the Ionic grid system. The Ionic grid system is based on flexbox (http://www.w3.org/TR/css3-flexbox/), which is a CSS3 feature that is supported by all devices. Grids contain three components: grid, rows, and columns. Here are some important points to note about the grids system:

- `ion-grid` and `ion-col` have padding by default of 5px, but the `ion-row` has no padding and no margin.
- The `no-padding` attribute can be added for removing that padding.
- Twelve column layout is used, similar to what we have in bootstrap grid system.
- `column` attribute is used in `ion-col` for setting width, such as `<ion-col col-3>` and the default, is col-12. Also, by default columns take up equal width inside of a row for all devices and screens.
- There are five grid tiers: `col-xs`, `col-sm`, `col-md`, `col-lg`, and `col-xl` which correspond to different screen sizes.

- offset is an attribute added to set offset which increases the margin left by `*` columns, such as `<ion-col offset-1>`
- We can align columns in the row such as `align-self-center` which will align column in the center. For more attributes details check `https://ionicframework .com/docs/api/components/grid/Col/`.
- Similarly we have `row` attributes and by default columns will stretch to fill entire height of row. For more attributes details check `https://ionicframework.com/d ocs/api/components/grid/Row/`.

We have to use grids while we design different layouts in our application. The simplest layout can be built using `width` and `offset` attributes. We have built a layout for our home page of the e-commerce app in which we have two cards in a row:

```
<ion-grid no-padding>
 <ion-row>
  <ion-col no-padding col-6 col-md-3 *ngFor="let
  product
  of products;">
  <ion-card>
   <img [src]="product.imageUrl"/>
   <ion-card-content>
    <ion-card-title no-padding>
     {{product.productName}}
    </ion-card-title>
   <ion-row no-padding class="center">
   <ion-col>
     <b>{{product.price | currency }}   </b><span
         class="discount">{{product.listPrice |
         currency}}</span>
   </ion-col>
   </ion-row>
   </ion-card-content>
  </ion-card>
 </ion-col>
</ion-row></ion-grid>
```

With the new grid system it make life of developer really easy. Previously, just for aligning two cards in a row and iterate it we had to use multiple `*ngFor` and iterate it on `row` and `col`.

Now, this grid system is highly inspired by the bootstrap grid system which is really familiar to all of us, So we have two cards in a `row` on a small screen and for medium and above we will have four cards in a `row`:

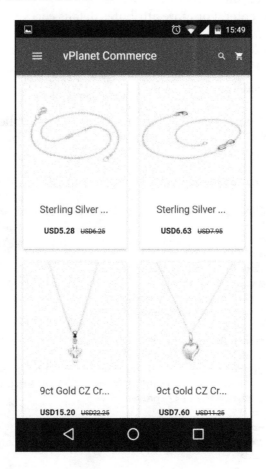

Custom pipes

Pipes are similar to filters in Angular 1, which take an input and transform the data before displaying it within the template. There are some sets of predefined pipes available, which we can directly use in our `template` files, such as date and currency. Other than this, we can also create custom pipes according to our requirements. In our application we are implementing an `Ordinalpipe`, which takes a number as input and accordingly outputs its ordinal number; for 1 the output will be 1st:

```
// src/filters/ordinal.ts

import { Pipe, PipeTransform } from '@angular/core';

@Pipe({name: 'ordinal'})

export class OrdinalPipe implements PipeTransform {
  transform(value: number, args: string[]): any {

    if (!value) return value;

    if(isNaN(value) || value < 1) {
      return value;
    } else {
      var lastDigit = value % 10;

      if(lastDigit === 1) {
        return value + 'st'
      } else if(lastDigit === 2) {
        return value + 'nd'
      } else if (lastDigit === 3) {
        return value + 'rd'
      } else if (lastDigit > 3) {
        return value + 'th'
      }

    }
  }
}

// using pipe in template

<p> {{ data.value | ordinal }} </p>
```

Other than creating a pipe and using it, you have to declare the pipe inside src/app/app.module.ts, as we declare for other modules:

```
import { OrdinalPipe } from '../filters/ordinal';

@NgModule({

  declarations: [
    .
    .
    .
    OrdinalPipe,
    Settings
  ]
```

```
})
```

This is a short version of what the `app.module.ts` file looks like; you can find the entire code in the file. Other than simple usage of pipes, we can do chaining in pipes as follows:

```
<p>{{ data.created_at | date | uppercase }}</p>
```

Navigation and view lifecycle

Navigation inside Ionic has been a debatable topic from the start. For developers coming from Ionic 1 and having a habit of using a UI router for navigation, their first concern was how to traverse through the application and declare different routes. I would say if we think the other way around towards the native world, we don't have routes declared like Ionic 1. Ionic 3 works a bit differently here; rather than using URLs to navigate, we are using `push` and `pop` operations for navigation.

Let's think of this as a stack where we push pages to move forward, and when we have to come back we pop that page. Initially when starting the application we declare the `rootPage` of the application, which is what you will see when you first open the application. We will first start with an example of how we used `rootPage` inside our e-commerce application. Initially, at the start of this chapter, we discussed how we set the initial `rootPage`, depending upon whether the user has seen the walkthrough page or not. Another similar example we will look at will be setting the `rootPage` after a successful login. After the user has logged in, we want it to clear the history stack, and no **back** button should be shown, as it will go back to the login page again:

```
// src/pages/login/login.ts

import { NavController, LoadingController, ToastController, Events } from
'ionic-angular';

constructor(public navCtrl: NavController,
    public loadingCtrl: LoadingController,
    public auth: Auth,
    public userService: UserService,
    public toastCtrl: ToastController,
    public events: Events) { }

this.auth.login('basic', this.login).then((result) => {
    // user is now registered
    this.navCtrl.setRoot(HomePage);
    this.events.publish('user:login');
    loading.dismiss();
    this.showToast(undefined);
}, (err: IDetailedError<string[]>) => {
```

```
        console.log(err);
        loading.dismiss();
        this.showToast(err)
});
```

Here we have cleared the navigation stack by calling the `setRoot` function. The `NavController` class here is the base for navigation, which handles a lot of things such as routing, history, route guards, and lifecycle events. In `Chapter 3`, *Ionic Native and Plugins,* we will also be covering Ionic deeplinks, which help us traverse directly to a URL. This will also help while adding support for PWA, as this will enable routing via unique SEO friendly URLs.

In our application, we used push operations multiple times and also set a new `rootPage`. In our application, the **cart** icon on the header is calling the `push` method in `NavController`:

```
<ion-buttons end>
    <button ion-button (click)="goToCart()">
        <ion-icon ios="ios-cart" md="md-cart"></ion-icon>
    </button>
 </ion-buttons>

// src/pages/categories/categories.ts

import { NavController, PopoverController } from 'ionic-angular';
import { ShowcartPage } from '../showcart/showcart';

export class CategoriesPage {
    constructor(public navCtrl: NavController) {}

    goToCart() {
        this.navCtrl.push(ShowcartPage);
    }
}
```

Here you can see we have pushed the `ShowcartPage` component. We can also pass data on to the next page as the next parameter to the `push` function and fetch it on the next page using the `navParams` class and its `get` method.

Navigating from overlay components

Recently, I faced a problem while navigating from a popover to a page. Every time I click on the link inside popover and directly push the next component, it freezes the application.

We will be implementing the popover component and inside that we will traverse to
SettingsPage and SupportPage:

```
// src/pages/popover/popover.ts

import { Component } from '@angular/core';
import { ViewController, NavController, App, ModalController } from 'ionic-
angular';

import { SupportPage } from '../support/support';
import { SettingsPage } from '../settings/settings';


@Component({
  template: &grave;
    <ion-list>
      <button ion-item
 (click)="goToPage('SettingsPage')">Settings</button>
      <button ion-item
 (click)="goToPage('SupportPage')">Support</button>
      <button ion-item
 (click)="close('http://ionicframework.com/docs/v2/getting          -
started')">Learn Ionic</button>
    </ion-list>
  &grave;
})
export class PopoverPage {

  classes: any;

  constructor(
    public viewCtrl: ViewController,
    public navCtrl: NavController,
    public app: App,
    public modalCtrl: ModalController
  ) {

    this.classes  = {
      'SettingsPage': SettingsPage,
      'SupportPage': SupportPage
    }


  }

  goToPage(page:any) {
    // Navigting from Overlay Components
    // https://ionicframework.com/docs/v2/api/navigation/NavController/
    // If you simply push it will freeze the UI
```

```
    this.viewCtrl.dismiss();
    this.app.getRootNav().push(this.classes[page]);
  }

  close(url: string) {
    window.open(url, '_blank');
    this.viewCtrl.dismiss();
  }
}
```

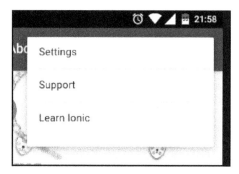

As you can see, rather than directly pushing the `Page` component, we first get a reference of `rootNav` in our app, using the `getRootNav()` method.

Lifecycle events and Nav Guards

Lifecycle events have been there in Ionic 1, and Native platforms also have a well structured lifecycle, which allows us to perform actions during various stages of navigation.
There are six Page events and two Nav Guards available:

- `ionViewDidLoad`: When the page has loaded
- `ionViewWillEnter`: When the page is about to enter
- `ionViewDidEnter`: When the page is fully entered
- `ionViewWillLeave`: When the page is about to leave
- `ionViewDidLeave`: When the page has finished leaving
- `ionViewWillUnload`: When the page is about to be destroyed
- `ionViewCanEnter`: Runs before the view can enter

- `ionViewCanLeave`: Runs before the view can exit

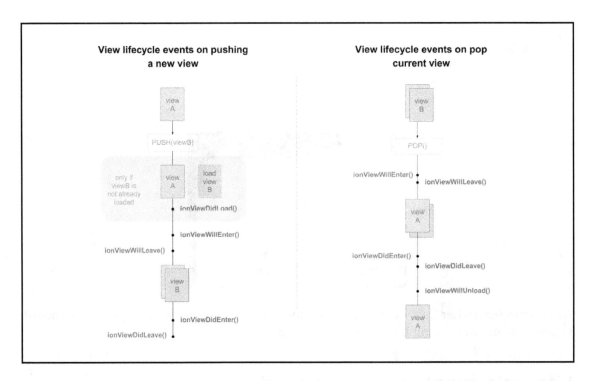

`ionViewCanEnter` and `ionViewCanLeaeve` are two Nav Guards that help control views leaving and entering. One such example is if we are filling up a form and accidentally click some button or menu item, it will immediately clear the data that the user has filled. So to secure the data we can apply `ionViewCanLeave`, like we did on our support page:

```
// src/pages/support/support.ts

// If the user enters text in the support question and
   then navigates
  // without submitting first, ask if they meant to leave
     the page
  ionViewCanLeave(): boolean | Promise<boolean> {
  // If the support message is empty we should just
    navigate
  if (!this.supportMessage ||
  this.supportMessage.trim().length === 0) {
    return true;
  }

  return new Promise((resolve: any, reject: any) => {
```

```
    let alert = this.alertCtrl.create({
      title: 'Leave this page?',
      message: 'Are you sure you want to leave this
      page? Your support message will not be submitted.'
    });
    alert.addButton({ text: 'Stay', handler: reject });
    alert.addButton({ text: 'Leave', role: 'cancel',
    handler: resolve });

    alert.present();
  });
}
```

Navigation is an important component for any application we build, so my personal advice would be to go through and understand all important topics big or small for the navigation component. One mistake I did initially was not going through the entire documentation, which almost wasted my entire day because of the freeze issue that I mentioned previously. You can go through the Navigation APIs

at https://ionicframework.com/docs/v2/api/navigation/NavController/.

Internationalization and localization

When we have to build a global application that touches the lives of people from different corners of the world, then we have to think about many aspects such as internationalization and localization. We will be integrating this inside our e-commerce app so that we can add multiple language support.

We will start by installing the `ng2-translate` npm module:

```
npm install ng2-translate --save
```

Next, we will import the modules and methods used inside `app.module.ts`:

```
// /src/app/app.module.ts
import { TranslateModule, TranslateLoader, TranslateStaticLoader } from
'ng2-translate/ng2-translate';

export function createTranslateLoader(http: Http) {
    return new TranslateStaticLoader(http, './assets/i18n', '.json');
}

@NgModule({
    imports: [
      IonicModule.forRoot(vPlanetApp),
      TranslateModule.forRoot({
        provide: TranslateLoader,
        useFactory: createTranslateLoader,
        deps: [Http]
      }),
      CloudModule.forRoot(cloudSettings)
    ]
})

export class AppModule {}
```

We have created a separate function, `createTranslateLoader`, in which we define the `i18n/` folder containing JSON files for each locale. We are now ready to use the module and we now initiate `TranslateService` inside the `app.component.ts` to which locale to use:

```
import {TranslateService} from 'ng2-translate/ng2-translate';

@Component({
  templateUrl: 'app.template.html',
  providers: [TranslateService]
})
```

```
export class vPlanetApp {

    constructor(
        public translate: TranslateService
    ) {

    this.initializeTranslateServiceConfig();


    }

    // We are using the user default language
    // and set that default locale

    initializeTranslateServiceConfig() {
        var userLang = navigator.language.split('-')[0];
        userLang = /(en|es)/gi.test(userLang) ? userLang :
        'en';
        this.translate.setDefaultLang('en');
        this.translate.use(userLang);
    }

}
```

Now we can start using it inside our template similar to how we used pipes. Because we have side menu list items that we iterate on inside our e-commerce application, we will be initially implementing translation on the sidemenu:

```
<ion-list>
    <ion-list-header>
      {{'navigate' | translate}}
    </ion-list-header>
    <button ion-item menuClose *ngFor="let p of
     appPages" (click)="openPage(p)">
      <ion-icon item-left [name]="p.icon"
       [color]="isActive(p)"></ion-icon>
      {{ p.title | translate }}
    </button>
</ion-list>
```

Now as soon as the locale is set it will pick up the exact labels from the JSON files, which we have inside the /assets/i18n folder. Inside this application we have implemented en and es locale only for now, but we can easily expand it by adding more locale.json files for each language that we want to support in our e-commerce application:

```
// es.json
{
```

```
    "menu": "Menú",
    "home": "Casa",
    "navigate":"Navegar",
    "categories":"Categorías",
    "cart":"Carrito de compras",
    "wishlist":"Lista de deseos",
    "about": "Acerca de",
    "categories":"Categorías",
    "login": "Iniciar sesión",
    "logout":"Cerrar sesión",
    "account":"Cuenta",
    "signup":"Regístrate",
    "popular":"Productos Populares",
    "settings":"Ajustes",
    "feedback":"Realimentación"
}
```

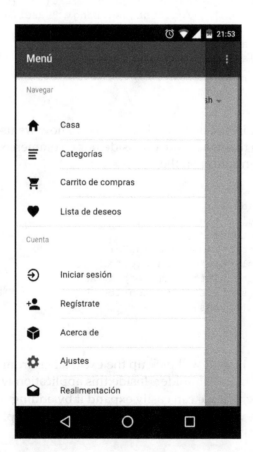

The preceding is an example `es.json` inside which we have placed different labels for which we will be translating the content. Next, if you want to translate any JavaScript value inside our `TypeScript` file then we can use `translateService` and its get method, which then subscribes to the observable and returns the translated string:

```
translateService.get('HELLO').subscribe(
  value => {
   // value returned is translated
   console.log(value);
  }
)
```

Setting up and translating the module initially might take, some time as developers might miss some points while setting it up due to multiple moving parts. Adding internationalization to your application will make a huge difference and will broaden the audience you are able to attract; this will help you gain a better rating on the Play Store.

ItemSliding and pull-to-refresh

Animations and gestures in modern day applications are a must and they enhance the user experience while using the application features. You will find the `ItemSliding` feature in most iOS applications, and pull-to-refresh is commonly used to fetch the latest data, and as in the Facebook application, you will use this to fetch the latest feeds.
`ItemSliding` is a swipe-able list that can be swiped towards the left or right to reveal action buttons. We will be using a swipe-able list on our wishlist page, where we have products action to delete from the list or add to cart:

```
//   src/pages/wishlist/wishlist.ts
<ion-content>
   <ion-list>
    <ion-list-header>
      4 Items in your Wishlist (Swipe for options).
    </ion-list-header>

    <ion-item-sliding *ngFor="let product of products">
      <ion-item>
        <ion-avatar item-left>
            <img [src]="product.image">
        </ion-avatar>
        <h2>{{product.name}}</h2>
        <h3>{{product.price | currency}}</h3>
        <p>{{product.rating}} stars and 420
    reviews</p>
      </ion-item>
```

```
      <ion-item-options side="right">
         <button danger>
            <ion-icon name="trash"></ion-icon>
            Remove
         </button>
         <button secondary>
            <ion-icon name="cart"></ion-icon>
            Add Cart
         </button>
      </ion-item-options>
   </ion-item-sliding>
 </ion-list>
</ion-content>
```

Let's look at some of the properties available on `temSliding`:

- Swipe direction can be right or left; both reveal the action buttons. We can use `<ion-item-options side="right">` for right swipe options
- `IonDrag` and `ionSwap` events can be used to listen or call some function during swiping of the list item
- Icon placement can be on the top or towards the left of the button text, such as `<ion-item-options icon-left>`
- The `ItemSliding` class can also be imported to access the `close()` method

As you can see, the preceding screenshot shows how the `itemSliding` list will look. Pull to refresh is another component that is mostly used when we need to refresh some existing content and fetch new data from backend servers. The `refresher` module detects the vertical swipe and display **loader** icon while calling the `refresh` function. We don't have the exact use case for using it inside our application, but it's really easy to implement the `refresher`:

```
<ion-content>
  <ion-refresher (ionRefresh)="updateFeed($event)">
    <ion-refresher-content
      pullingText="Updating Feed"
      refreshingSpinner="circles"
      refreshingText="Updating...">
    </ion-refresher-content>
  </ion-refresher>
  <h4>User Feed</h4>
</ion-content>
```

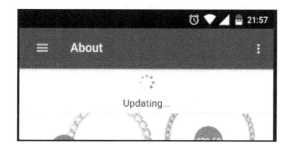

The `ion-refresher` component has the `ionRefresh` attribute, which is called when the `refresher` is pulled vertically to some extent. After that, as you can see inside the `ion-refresher-content`, there are multiple sets of attributes that are actually acting as configuration settings for the `refresher` having multiple options:

```
import { Http } from '@angular/http';

export class FeedPage {
    constructor(public http: Http) {}

    updateFeed(refresher) {

        // Calling a web service to fetch new feeds
          result
        this.http.get('path/to/data.json')
            .map(res => res.json())
            .subscribe(data => {
                // After the callback we will
```

```
                         complete the refresher
                 // and update the data inside the
                 view.
                        refresher.complete();
            });
    }
}
```

After we call the `ionRrefresh` function we need to close the refresher process once the data is retrieved from the server. The `complete()` method is called when our `async` request from the server is completed. We need to make sure we call this method otherwise the **loader** icon will not disappear.

Virtual scroll and infinite scroll

Virtual scroll and infinite scroll are two important components that are again widely used across different applications to cater for the high number of items in lists. Virtual scroll is implementing a list of items that can be too long to render at once, for example contact lists with thousands of contacts. If we try to load the entire list and render that, it will hinder the performance of our application and you will find lag while scrolling down the list. Let's look into how we can implement virtual scroll in our `products` page:

```html
// src/pages/products/products.html
<ion-content class="products">
    <ion-list>
                <ion-item>
                  <ion-note item-left>
                  Duplicate products for VirtualScroll demo
                </ion-note>
            </ion-item>
    </ion-list>
    <ion-grid [virtualScroll]="products"
     approxItemHeight="40px">
      <ion-row *virtualItem="let product"
        (click)="goToProductDetail(product)">
        <ion-col width-33>
          // Important to use ion-img rather than img tag
          // this will lazy load the images as we scroll
             downwards
          // Also, image sizes should not vary after its
             loaded
          <ion-img class="v-img" [src]="product.image">
         </ion-img>
        </ion-col>
        <ion-col width-67>
```

```
        <p>{{product.name}}</p>
        <ion-badge item-right>4.1</ion-badge> 340 ratings
            <p class="price"> {{ product.price |
        currency: 'INR' }} </p>
        </ion-col>
    </ion-row>
  </ion-grid>

</ion-content>
```

Some important points to note while working with the virtual scroll component are that we should not do any DOM manipulations or changing of datasets after the list is loaded, as this will require the entire list to reset or refresh again, which is an expensive operation on memory. Approximate widths and heights of the items inside a list are necessary to specify for virtual scroll to work smoothly. The default is 40px height, although it's not necessary to give the exact number, but it will be better if it is close to the exact height.

Infinite scroll or endless scrolling is where the content of the page is appended dynamically after a user scrolls a specified distance. In infinite scrolling, a function is called to asynchronously fetch new data and append it to the list:

```
<ion-content>

  <ion-list>
    <ion-item *ngFor="let product of products">
      {{product.name}}</ion-item>
  </ion-list>

  <ion-infinite-scroll
    (ionInfinite)="getNewProducts($event)">
    <ion-infinite-scroll-content></ion-infinite-scroll-
      content>
  </ion-infinite-scroll>

</ion-content>
```

The `ionInfinite` attribute is used to call the `getNewProducts()` function, which will eventually fetch the new data:

```
export class ProductsPage {

  constructor() {}

  getNewProducts(infiniteScroll) {
    // Call a http function or service
    // fetches new data and append to the products array
        list
```

```
      // Make sure to call complete method to close the
         loader icon
      infiniteScroll.complete();
   }

}
```

As you can see, both virtual and infinite scroll are almost the same with respect to the template and its methods. You can find similar attributes in both components, where you can declare `loadingText` and `loadingSpinner`. If required we can further customize both virtual and infinite scroll design with custom SVG and CSS animations.

Ionic rating

Ionic rating is a third-party module for visualizing the rating bar. We will be using this module in our application in the product-detail page. This module is available as an NPM module, so it will be really easy and quick to integrate:

```
$ npm install --save ionic2-rating
```

```
// src/app/app.module.ts

import { Ionic2RatingModule } from 'ionic2-rating';

imports: [
    IonicModule.forRoot(vPlanetApp),
    Ionic2RatingModule,
    TranslateModule.forRoot({
      provide: TranslateLoader,
      useFactory: createTranslateLoader,
      deps: [Http]
    }),
    CloudModule.forRoot(cloudSettings)
  ],

// src/pages/product-detail/product-detail.html
<rating [(ngModel)]="rate" readOnly="false" max="5"></rating>
```

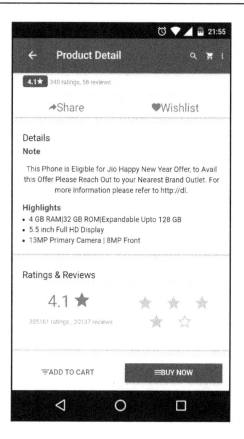

The Ionic community has evolved a lot in the last three years; there are multiple third-party components available on GitHub, which you can smartly use inside your application the same as we used the ionic-rating plugin for our e-commerce application. Some of the components are:

- Ionic rating (https://github.com/andrucz/ionic2-rating)
- Image cache and loader (https://github.com/zyramedia/ionic-image-loader)
- Mandatory install (https://github.com/NextFaze/ionic-manup)
- Geofencing (https://github.com/tsubik/ionic2-geofence)
- Accordion (https://github.com/mahmoudissmail/ionic2Accordion)
- Calendar (https://github.com/twinssbc/Ionic2-Calendar)

The best way to find these awesome pieces of code is by searching on GitHub and looking for the top starred ionic projects. I have been using this hack for learning what developers are silently doing or working on Ionic. All these projects are MIT licensed most of the time so you can freely use them inside your application.

Lazy Loading with Ionic 3

Lazy Loading is much-awaited feature which recently released with Ionic 3. Lazy Loading will improve our applications startup time, reduce the bundle size and removes the hassle of including same component path in every Class which eases the process of setting up Ionic routing. Let's look into what changes are required and how your application will improve. Also, in Chapter 5, *Authentication, Authorization, and Security* the application demonstration will have lazy Loading implemented inside it.

Currently, lazy Loading component have some known bugs which Ionic team is actively working upon. Also, setting this will override DeepLinkConfig defined in IonicModule.forRoot. We will be introducing IonicPageModule in each component and new module.ts file in each page component. Make sure you actively follow Ionic blog and any new releases which can come up with further bugs fixes and new features.

Previously, we import all the pages in one go inside our app.module.ts and now we just load our app.component.ts initially. So, all these components like HomePage will be removed from declarations and entry components. Next inside our src/pages/home directory we have introduced another file home.module.ts similar to what we have app.module.ts.

Now if you will generate a page using Ionic CLI you will find four files created:

```
home.html

home.module.ts

home.scss

home.ts

// src/pages/home/home.ts

import { Component } from '@angular/core';

import { IonicPage, NavController } from 'ionic-angular';

@IonicPage()

@Component({

  selector: 'page-home',

  templateUrl: 'home.html',

})
```

```
export class HomePage {

    constructor(public navCtrl: NavController)

  moveTo(data:any) {

    this.navCtrl.push("ContactPage", data);

  }

  }
```

Also, now if we want to Push or Pop to any other `Component` we don't require to import it. We can directly handle that as string, during build process deeplinks will be generated that know how to handle that string. The string is actually a reference to the `name` property of the `@IonicPage` decorator, which defaults to the class name as a string. If we change that name property to something else, we'll also need to update the reference we use elsewhere. You can find more about `IonicPage` and supported field here (`http://ionicframework.c om/docs/v2/nightly/api/navigation/IonicPage/`)

Summary

Ionic components and related resources are the backbone of any Ionic application. With intelligent use of components and available third-party components, we can get better performance and make our application user friendly. Choosing the right component for the right place is tricky, and the most interesting part is we have to think from the perspective of the end user and the end user's experience with our application. We have to make sure that the user does not feel as if they are in a different world and away from the native experience, which usually happens. We can cover as many important components and important points as we want in this chapter, but what is suggested is to stay updated in the community and look for available resources that you can utilize easily for your application. Next, we will be moving towards the integration of plugins within our application. As almost every application uses one plugin or another, it's very critical for production applications to integrate plugins correctly. Camera, GPS, filesystem, and network plugins are commonly used plugins. We will work on integrating these commonly used plugins and also some complex plugins that many developers have a problem with while integrating.

The Ionic team have already launched an awesome project, Ionic Native, which eases the process of using Cordova plugins. Ionic Native is a wrapper library written over many of the plugins, so we can easily integrate it inside Angular-based ionic applications.

3
Ionic Native and Plugins

We have now gone through some advanced components and APIs in Ionic. Another important part of building hybrid mobile applications are Cordova plugins. Plugins create a bridge between Native APIs and JavaScript, which help us to use Native APIs such as camera, GPS, files, and so on. Plugins are comprised of platform-specific Native library code and a JavaScript interface to call respective Native platform methods. Ionic is dependent on Cordova to build, debug, and release applications. We will be covering almost everything regarding Cordova plugins, Ionic Native, and some common and advanced plugins in this chapter. Initially, the Ionic team came up with ngCordova, which was based upon Angular 1.x, but moving forward, we have Ionic Native, which supports both Ionic 1, 2, and 3 applications. ngCordova was bound to Ionic 1 and Ionic Native is framework agnostic:

- Cordova and how plugins work
- Ionic Native
- Commonly used Cordova plugins (camera, GPS, contacts)
- Building a music player
- Cordova versus phonegap

Cordova and how plugins work

Cordova is a platform for building Native mobile applications within a single codebase in HTML, CSS, and JavaScript. Cordova provides support to multiple platforms and plugins. Cordova plugins allow Cordova web view to communicate with Native platforms and then our Ionic application runs inside the web view, so this way we have access to the platform APIs, such as camera, geolocation, and so on.

What is Cordova?

As mentioned about Cordova in the official line:

> *Apache Cordova is a project maintained by the Apache organization. Cordova is a set of device APIs that allow a mobile app developer to access native device functions such as the camera or accelerometer from JavaScript. Combined with frameworks such as Ionic, this allows smartphone apps to be developed with just HTML, CSS, and JavaScript.*

As we all might have used Cordova in our application and also know some common plugins that Cordova supports officially, I will lay down some of the common points that will help us understand Cordova:

- Cordova is an application container
- Cordova helps build Native applications for each platform
- Cordova provides access to Native device APIs such as camera, contacts, GPS, and so on
- Custom plugins can also be written and supported
- Cordova is the engine behind Adobe phonegap
- Cordova is open source and free to use

As mentioned previously, Cordova officially supports multiple platforms and plugins. We can see that a lot of new plugins are coming, such as for fingerprints, hotspot, and so on, which we can utilize in building our advanced hybrid applications. The majority of the third-party plugins support iOS and Android platforms as these platforms have a major share in the device market. Recently, growing technology such as Bluetooth beacons are also supported by some third-party Cordova plugins. Later on in this book we will build an application and work with this technology.

Building and using Cordova plugins

Developers use CLI commands to build or add plugins for Cordova in their project. You can also publish your plugins and share them with the community with plugman. We will also quickly go through how we can build and publish our plugins.

Plugin specifications

All the plugin repositories must feature a top-level `plugins.xml` manifest file. This file defines the plugin, various platforms it supports, and its structure. There are multiple specifications in this file, some of these are mentioned here:

- `name`: Human readable name
- `description`: Description about the plugin
- `author`: Author name
- `keywords`: Keywords that describe the plugin
- `licence`: License of the plugin
- `js-module`: Including JavaScript files
- `platform`: For including multiple platform and associated code
- `source-file`: Identifies executable source code
- `dependencies`: For adding dependencies on other plugins

The following is an example `plugins.xml` file for a simple Cordova plugin:

```xml
<?xml version="1.0" encoding="utf-8"?>
<plugin xmlns="http://www.phonegap.com/ns/plugins/1.0"
        id="cordova-eddyblue"
        version="1.0.0">

<name>Custom Plugin</name>
<description> Cordova custom bluetooth plugin
    </decription>
<author>Gaurav Saini</author>
<license>Apache 2.0</license>
<keywords>Eddystone, Beacons</keywords>


<engines>
  <engine name="cordova" version=">=3.4.0"/>   <!--
      Needs cordova version greater than 3.4.0 -->
</engines>

<!-- Javascript module and file path -->
<js-module src="www/eddyblue.js" name="eddyblue">
  <clobbers target="eddyblue" />
</js-module>

<!-- android platform -->
<platform name="android">

  <config-file target="res/xml/config.xml"
        parent="/*">
    <feature name="Scan">
      <param name="android-package"
          value="com.eddyblue.plugin.scan"/>
    </feature>
```

```
      </config-file>

      <source-file src="src/android/Eddyblue.java"
          target-dir="src/com/eddyblue/plugin/"/>
    </platform>

    <!-- ios platform -->
    <platform name="ios">

      <config-file target="config.xml" parent="/widget">
        <feature name="Scan">
          <param name="ios-package" value="HWPEddyblue"
            />
        </feature>
      </config-file>

      <header-file src="src/ios/HWPEddyblue.h" target-
          dir="Eddyblue"/>
      <source-file src="src/ios/HWPEddyblue.m" target-
          dir="Eddyblue"/>
    </platform>

  </plugin>
```

Plugman

Plugman is a utility for validating, installing, and publishing your custom build plugin for Cordova. Plugman can be installed via the NPM command:

```
$ npm install -g plugman
```

You need to test and validate your plugin in your project before you can publish it. Using `plugman install` you can install the plugin for a specific platform, as mentioned previously:

```
$ plugman install --platform android --project /path/to/project/www --
plugin /path/to/custom-plugin
```

We will not go deep into `plugman`, but you can perform functions such as installing, uninstalling, search registry, and getting information via `plugman`.

JavaScript interface

Cordova plugins have a front facing interface, which is a JavaScript interface and we call its `function` from our application. The `cordova.exec` call communicates with the Native platform. The following is an example syntax:

```
module.exports = {
scan: function (name, successCallback,
    errorCallback) {
cordova.exec(successCallback, errorCallback,
    "Scan", "start", [name]);
}
};
```

Native platform interface

As we define the JavaScript interface of our plugin, we also need a Native implementation for each platform that the plugin supports. The following is a sample Java implementation for the Native Android platform:

```
package com.eddyblue.plugin;

import org.apache.cordova.*;
import org.json.JSONArray;
import org.json.JSONException;

public class Scan extends CordovaPlugin {

    @Override
    public boolean execute(String action, JSONArray
        data, CallbackContext callbackContext) throws
        JSONException {

        if (action.equals("start")) {

            String message = "data to be sent";
            callbackContext.success(message);

            return true;

        } else {

            return false;

        }
    }
```

```
}
```

We have now briefly covered how plugins work. You can now easily share your plugin with the community and add it to the registry using plugman. For more details about plugin development, you can refer to the Cordova official documentation, which can guide you further.

Ionic Native

Developers working on hybrid applications have been using ngCordova with Ionic 1 for some time. Now with Ionic 3, which is built with ES6 and TypeScript, we need to have a wrapper library for ES6 or TypeScript, which is exactly what Ionic Native is. It's a collection of ES5/ES6/TypeScript wrappers built on top of Cordova APIs, which makes it convenient to integrate into Ionic applications. The Ionic team started this project as a feature request from a community member. It started as an experiment and it gained a lot of support due to the fact that integrated plugins were always a pain when it comes to lack of documentation and confusing APIs. These Angularj wrappers simplified the process of integration and fit with the Angular syntax also. Now there are around 120+ Cordova plugins supported in Ionic Native and still counting.

Installing Ionic Native

We can add Ionic Native to any existing Ionic project. You will find `ionic-native` by default in any starter project you will start with Ionic 3. If you somehow have to install it then the recommended way is via the NPM package:

```
$ npm install @ionic-native/core --save
```

If you are using ES6/TypeScript for building Ionic 3 apps, you do not need any scripts tags to be included in the `index.html` file. Just import the specific `ionic-native` package inside your file:

```
import { SplashScreen } from '@ionic-native/splash-screen'
import { StatusBar } from '@ionic-native/status-bar';
```

If you are using Ionic 1 then you need to add the `ionic.native.js` file inside the `index.html` file:

```
<script src="ionic.native.js"></script>
```

We have to wrap most of the plugin calls with the device ready event, so to check if your device has fully loaded the application and plugins are available, use the following code for the `Splashscreen` plugin:

```
constructor(platform: Platform, private
   splashScreen: SplashScreen, statusBar: StatusBar)
   {
     this.platform.ready().then(() => {
     // platform is ready to use available plugins.
     this.statusBar.styleDefault();
     this.splashScreen .hide();
   });
}
```

Testing plugins in browsers

One of the biggest issues currently with Cordova plugins is we can't test our app inside browsers. Ionic team have been concentrating on improving Ionic tools so that we can simulate most of our application's functionality directly in a browser. Still, for current development, there is a really nice Chrome plugin, Cordova mocks, which was created by Paolo Bernasconi, one of the core contributors to previous ngCordova projects. As soon as you install the plugin you can start testing your application as if you are testing it on your phone. Currently, there are around twelve plugins supported by this extension. This Chrome plugin can be effective in some cases and help you test in a browser for supported plugins.

Commonly used Cordova plugins

Whenever you start a new Ionic project and add a platform to it, you can see that there are some plugins that are installed by default. Ionic installs the following mentioned plugins when you add Android or iOS platforms:

- ionic-plugin-keyboard
- cordova-plugin-console
- cordova-plugin-device
- cordova-plugin-splashscreen
- cordova-plugin-whitelist
- cordova-plugin-statusbar

Let's get started with the previously mentioned plugins and see what they are used for.

Device plugin

Device plugin is really helpful in getting all the device related hardware and software information. Device objects are available in global scope, but only after the device is ready.

Ionic Native supports this plugin with devices and it has six static members to access data:

- `Device`: Returns Object with Cordova, model, platform, UUID, and version, for example, `Device.uuid`
- `cordova`: Returns string with specific Cordova versions
- `Model`: Returns string with the device's model name
- `platform`: Returns string with the operating system name
- `uuid`: Returns string with the device's **Universally Unique Identifier** (UUID)
- `version`: Returns string with the operating system version
- `isVirtual`: Checks whether the device is running on a simulator or not
- `serial`: Fetches the hardware serial number
- `manufacturer`: Fetches the device's manufacturer

```
import { Device } from '@ionic-native/device';

constructor(platform: Platform, private device:
 Device)
  {
   this.platform.ready().then(() => {
   console.log(this.device.version);
  });
 }
```

Splash screen plugin

The splash screen plugin is used to display and hide splash screen on the launch of the application. With Ionic applications you will get a default Ionic splash screen. You can easily change that with the Ionic CLI command Ionic resources after adding a new screen to the respective resources folder.

Splash screen is supported by Ionic Native and it has two methods:

- `this.splashScreen.show()`: Displays splash screen
- `this.splashScreen.hide()`: Hides splash screen

Whitelist plugin

This plugin implements a `whitelist` policy for network, navigating, and intent of the application web view on Cordova 4.0 and above versions. The `whitelist` plugin comes by default in the Ionic application when you start a new application, without this plugin you will not be able to access domains other than `file:///` URLs:

```
URL blocked by whitelist: http://maps.google.com
```

Network whitelisting

All the controls that the network requests (images, XHRs, and so on) are allowed to be made (via Cordova Native hooks). As now it comes by default, you will see in your `config.xml` a property, `<access origin>`, which allows all URLs. If you don't specify the access origin, only `file:///` URL requests are allowed:

```
<access origin="*" />  // Allow are URLs
```

Navigation whitelisting

This controls all the URLs the web view itself can be navigated to. Applies to top-level navigations only. In case we don't specify any specific URL, the default `file:///` is used for navigating web views:

```
<!-- Allow links to google.com →

<allow-navigation href="http://google.com/*" />

<!-- Wildcards are allowed for the protocol, as a
     prefix to the host, or as a suffix to the path
     example maps.google.com, mail.inbox.com -->

<allow-navigation href="*://*.google.com/*" />
```

Intent whitelisting

Intent whitelisting controls the URLs that the app is allowed to ask the system to open. By default, no external URLs are allowed. This whitelist does not apply to plugins, only hyperlinks and calls to `window.open()`. Also, you can allow specific intent, which can be accessed with the `<allow-intent>` property:

```
<allow-intent href="http://*/*" />
<allow-intent href="https://*/*" />
<allow-intent href="sms:*" />
<allow-intent href="tel:*" />

<!-- Allow all unrecognized URLs to open installed
    apps
 *NOT RECOMMENDED* -->
<allow-intent href="*" />
```

One important thing to note is you need to add the `Content-Security-Policy` meta tag inside `index.html` for Android to prevent showing this info message:

```
"No Content-Security-Policy meta tag found. Please add
one when using the cordova-plugin-whitelist plugin."
```

Some example `<meta>` tags are mentioned here, which we can utilize in our application:

```
<!-- Allow requests to gauravsaini.me -->
<meta http-equiv="Content-Security-Policy"
    content="default-src 'self' gauravsaini.me">

<!-- Enable all requests, inline styles, and eval() --
    >
<meta http-equiv="Content-Security-Policy"
    content="default-src *; style-src 'self' 'unsafe-
    inline'; script-src 'self' 'unsafe-inline' 'unsafe-
    eval'">

<!-- Allow XHRs via https only -->
<meta http-equiv="Content-Security-Policy"
    content="default-src 'self' https:">

<!-- Allow iframe to https://cordova.apache.org/ -->
<meta http-equiv="Content-Security-Policy"
    content="default-src 'self'; frame-src 'self'
    https://maps.google.com">
```

Camera plugin

The camera plugin is one of the most used plugins and it provides an API for taking photos and choosing images from the gallery. Although `navigator.camera` is available globally and can be accessed from anywhere, it can only be accessed after device ready.

You can start using this plugin after installing the camera plugin. Ionic Native is already installed with NPM and you can directly import the camera plugin from the `ionic-native` package:

```
$ ionic cordova:plugin add cordova-plugin-camera
```

```
$ npm install @ionic-native/camera --save
```

There are two main methods in the camera plugin:

- `Camera.getPicture`

- `Camera.cleanup`

For taking photos, using the camera, or choosing from the image gallery we use the `Camera.getPicture` method. It opens the camera or gallery depending upon the `sourceType` mentioned in the options:

```
// src/pages/account/account.ts


import { Camera, CameraOptions } from '@ionic-
        native/camera';

export class AccountPage {
 username: string;
 account: string = "profile";
 base64Image: any = "http://www.gravatar.com/avatar?
 d=mm&s=140";

 constructor(public alertCtrl: AlertController,
 public nav: NavController, public userService:
 UserService, private camera: Camera) {

 }

 updatePicture() {
 let options =const options: CameraOptions = {
 quality:100, // Specify quality in number 0-100
 destinationType:
```

```
Camerathis.camera.DestinationType.DATA_URL,
sourceType:
Camerathis.camera.PictureSourceType.CAMERA, //
camera or gallery
allowEdit: true,
encodingType: Camerathis.camera.EncodingType.JPEG,
targetWidth: 100,
targetHeight: 100,
saveToPhotoAlbum: true,
correctOrientation:true,
cameraDirection: 0// BACK 0, FRONT 1
};

Camerathis.camera.getPicture(options).then((imageDa
ta) => {
console.log(imageData);
this.base64Image = 'data:image/jpeg;base64,' + imageData;

}, (err) => {
// Handle error
 });
  console.log('Clicked to update picture');
 }

}
// src/pages/account/account.html


<img alt="profileImage" [src]="base64Image">
```

For fetching the image from the gallery you can change the sourceType to photolibrary. There are multiple options passed to the getPicture method. Let's check all the options that we can use:

- quality: Quality of the saved image, which can be in the range of 0-100 or default 50.
- destinationType: DATA_URL(base64 encoded), FILE_URL(file URI), and NATIVE_URL.
- sourceType: photolibrary, camera, and savephotoalbum.
- allowEdit: Allows simple editing of images before selection.
- encodingType: JPEG or PNG.
- targetWidth: Width in pixels to scale images. Aspect ratio remains constant. (Number)

- targetHeight: Height in pixels to scale image. Aspect ratio remains constant. (Number)
- mediaType: Type of media picture, video, or all media to select and it only works when PictureSourceType is photolibrary or savephotoalbum.
- correctOrientation: Rotate the image to correct the orientation of the device during capture. (Boolean)
- saveToPhotoAlbum: Save the image to the photo album on the device after capture. (Boolean)
- popoverOptions: iOS-only options that specify popover location in iPad.
- cameraDirection: Choose the camera to use front or back, default is back.

In case of destinationType FILE_URL, for removing photos taken by the camera from temporary storage, use the following code:

```
this.camera.cleanup().then(...); // only for FILE_URI
```

Photos selected from the gallery via changing sourceType are not downscaled to a lower quality, even if a quality parameter is specified. To avoid common memory problems, set Camera.destinationType to FILE_URI rather than DATA_URL.

Geolocation plugin

The geolocation plugin leverages the device's **Global Positioning System (GPS)** and fetches location information such as latitude and longitude. There are other sources also from which this plugin tries to get a device's location including Wi-Fi network and mobile network IP address, although this will not have high accuracy. This plugin is officially supported by Apache Cordova and it has a global accessible object navigator.geolocation, which again needs a deviceready event available for using its features.

Let's get started by installing the plugin via:

```
$ ionic cordova:plugin add cordova-plugin-geolocation
$ npm install @ionic-native/geolocation --save
```

There are mainly three methods available in this plugin:

- `getCurrentPosition`
- `watchPosition`
- `clearWatch`

`getCurrentPosition` returns the position of the current position:

```
let options = {
timeout :3000,
enableHighAccuracy: true,
maximumAge : 1000
};

this.gGeolocation.getCurrentPosition(options).then(( position)
=> {
// on success
let cordinates = position.coords.latitude
+","+position.coords.longitude
}).catch((err) => {
// on error
console.log(err.message);
});
```

As you can see from the preceding example, we have passed options to the `getCurrentPosition()` function. Similar options are provided to the `watchPosition()` function. `WatchPosition()` watches the location changes and when the device retrieves a new location, callback executes with the position object having the latest location data:

```
let watchOptions = {
timeout : 3000,
enableHighAccuracy: false // may cause errors if true
};
let watch = this.Ggeolocation.watchPosition(watchOptions);

watch.subscribe((position) => {
let lat = position.coords.latitude
let long = position.coords.longitude
});

// Clearing the watch
watch.clearWatch()
```

Also, we can easily clear the watch method; this will stop watching for changes in the device's location. While using `getCurrentPosition` and `watchPosition` method we pass options to it, which have three properties:

- `timeout`: The maximum length of time (milliseconds) that is allowed to pass until the success or error callback. If a success callback is not evoked the error callback is passed the `TIMEOUT` error code. (Number).
- `enableHighAccuracy` : Provides a hint that the application needs the best possible results. By default, the device attempts to retrieve a position using network-based methods. Setting this property to `true` tells the framework to use more accurate methods, such as satellite positioning. (Boolean).
- `maximumAge`: Accepts a cached position whose age is no greater than the specified time in milliseconds. (Number).

You can build very interesting applications using the geolocation plugin such as finding the distance between two coordinates to check your nearby restaurants that are in a 5 km radius. The Haversine formula (`https://en.wikipedia.org/wiki/Haversine_formula`) helps us to find the distance between your current position and the destination through which you can include and exclude a resulting list of restaurants with specific radius.

Social sharing plugin

Sharing content with social media is a great tool for marketing your application and its content, which helps you gain new users. So having a share plugin will be a really fruitful feature for our applications. You can share images, text, messages via Facebook, Twitter, e-mail, SMS, WhatsApp, and so on using this plugin. It uses the Native share plugin to share content and will automatically show all sharing options that you have in your device:

```
$ ionic cordova:plugin add cordova-plugin-x-
    socialsharing
$ npm install @ionic-native/social-sharing --save
```

`SocialSharing` has multiple methods that we can use in our application. The most commonly used method is `SocialSharing.share`, which opens the default platform dialog with all the available sharing applications:

```
// src/pages/about/about.ts

import { SocialSharing } from '@ionic-native/social-sharing';

export class AboutPage {
```

```
constructor(public popoverCtrl: PopoverController, private
socialSharing: SocialSharing) { }

shareViaFacebook() {
// recommended to use canShareVia before using shareVia
this.sSocialSharing.canShareVia("facebook", "Veloice:Intelligent
Business Telephony", "Real time Voice",
"http://veloice.com/images/banner.png",
"http://veloice.com").then(() => {
this.sSocialSharing.shareViaFacebook("Veloice:Intelligent Business
Telephony", "http://veloice.com/images/banner.png",
"http://veloice.com");
}).catch(() => {
alert("Cannot share on Facebook");
});
}

shareViaWhatsApp() {
// recommended to use canShareVia before using shareVia
this.sSocialSharing.canShareVia("whatsapp", "Veloice:Intelligent
Business Telephony", "Real time Voice",
"http://veloice.com/images/banner.png",
"http://veloice.com").then(() => {
this.sSocialSharing.shareViaWhatsApp("Veloice:Intelligent Business
Telephony", "http://veloice.com/images/banner.png",
"http://veloice.com");
}).catch(() => {
alert("Cannot share on Whatsapp");
});
}

shareViaEmail() {
// recommended to use canShareVia before using shareVia
this.sSocialSharing.canShareViaEmail().then(() => {
this.sSocialSharing.shareViaEmail(
'Veloice:Intelligent Business Telephony', // Message
'Veloice', // Email Subject
['toperson@xyz.com', 'tosecond@xyz.com'], // TO: must
 be null or an array
['cc@xyz.com'], // CC: must be null or an array
null, // BCC: must be null or an array
['http://veloice.com/images/banner.png'] // FILES: can
 be null, a string, or an array
);
}).catch(() => {
alert("Cannot share on Email");
});
}
```

```
share() {
this.sSocialSharing.share("Veloice:Intelligent
Business Telephony", "Real time Voice",
"http://veloice.com/images/banner.png", "http://veloice.com");
 }

}
```

```
// src/pages/about/about.html

<ion-fab right bottom>
 <button ion-fab><ion-icon name="help"></ion-icon></button>
 <ion-fab-list side="top">
 <button (click)="callNow()" ion-fab><ion-icon
  name="call"></ion-icon></button>
 <button (click)="shareViaEmail()" ion-fab><ion-icon
  name="mail"></ion-icon></button>
 <button (click)="shareViaFacebook()" ion-fab><ion-icon
  name="logo-facebook"></ion-icon></button>
 <button (click)="shareViaWhatsApp()" ion-fab><ion-icon
  name="logo-whatsapp"></ion-icon></button>
 <button (click)="share()" ion-fab><ion-icon
  name="share"></ion-icon></button>
 </ion-fab-list>
</ion-fab>
```

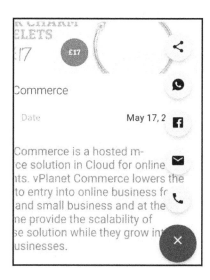

Previously listed are all the methods that the share plugin provides. You can see in the code that where we used the `shareViaFacebook` method, we wrapped it around the `canShareVia` method to pre check if the Facebook app is installed or not. Similarly, we can use it for Twitter, Instagram, and Whatsapp. Sharing via SMS and e-mail examples are also there, so we can directly share with those methods too. The share plugin also supports multiple images via array. The `SaveToPhotoAlbum` method is also available in the plugin core code, but until now Ionic Native did not have any wrapper for it. So you have to use it like this:

```
    window.plugins.socialsharing.saveToPhotoAlbum(
    ['http://veloice.com/icon.png','www/veloice.jpg'],
    onSuccess, // optional success function
    onError // optional error function
);
```

We have now covered some of the most used plugins. We will continue to look at more complex plugins that need detailed examples, such as Facebook and SQLite. Our main focus will be to continue to closely stick to Ionic Native and make full use of it.

Call number

The `call-number` plugin is used to call directly from your application. We have used it inside about page for calling the support number:

```
$ ionic cordova:plugin add call-number
$ npm install @ionic-native/call-number --save
```

Before using the plugin we have to install the Cordova plugin and NPM install specific package:

```
// src/pages/about/about.ts

import { CallNumber } from '@ionic-native/call-number';

constructor(private call: CallNumber) { }

callNow() {
 this.CallNumbercall.callNumber("+919989887765", true)
 .then(() => console.log('Launched dialer!'))
 .catch(() => console.log('Error launching dialer'));
 }
```

Text to speech

The `text-to-speech` plugin does what it says, it synthesizes text to voice. The plugin uses `AVSpeechSynthesizer` on iOS and the `android.speech.tts.TextToSpeech` package on Android. Let's get started with installing the `cordova` plugin and respective `npm` package:

```
$ ionic cordova:plugin add cordova-plugin-tts
$ npm install @ionic-native/text-to-speech --save
```

We have used it in our application to speak out a `push notification` message, whenever a `push` message is sent from the server:

```
// src/app/app.component.ts

import { TextToSpeech } from '@ionic-native/text-to-
   speech';
constructor(private tts: TextToSpeech) { }
push.on('notification', function(data) {
 TextToSpeechthis.tts.speak({text: data.message,
   locale: 'en-GB', rate: 0.75})
 .then(() => console.log('Success'))
 .catch((reason: any) => console.log(reason));
});
```

AppRate

The `AppRate` plugin is used to prompt a user to rate an app now, later, or never. Every application needs this plugin to get reviews and ratings from customers:

```
$ ionic cordova:plugin add cordova-plugin-apprate
$ npm install @ionic-native/app-rate --save
```

`AppRate` is really easy to integrate as all the logic for showing the prompt and other settings are managed by the plugin. We just need to set the preferences accordingly:

```
// src/app/app.component.ts

import { AppRate } from '@ionic-native/app-rate';

platformReady() {
 // Call any initial plugins when ready
 this.platform.ready().then(() => {
 Splashscreen.hide();
```

```
if (this.platform.is('cordova')) {
// App rate plugin
AppRatethis.appRate.preferences.customLocale = {
title: "Rate vPlanet Commerce",
message: "Would you mind taking a moment to rate
it? It won't take more than a minute. Thanks for
your support!",
cancelButtonLabel: "No, Thanks",
laterButtonLabel: "Remind Me Later",
rateButtonLabel: "Rate It Now"
};
AppRatethis.appRate.preferences.storeAppURL = {
android: 'market://details?
id=com.vplanetcommerce.demoionicpwa'
};
AppRatethis.appRate.promptForRating(false);

// Initiate Push
this.initPushNotification();
}
});
}
```

If we set `promptForRating()` to false then it will not prompt immediately . It will use the `usesUntilPrompt` preference, which counts the runs of an application before the dialog will be displayed. You can refer to official documentation for further options: https://ionicframework.com/docs/v2/native/app-rate/.

Google analytics

Analytics is key for any successful application; it helps understand which screens are most visited and which products users took interest in the most. We can track many different events that a user does while using the application:

```
$ ionic cordova:plugin add cordova-plugin-google-
   analytics
$ npm install @ionic-native/google-analytics --save
```

Let's start by initializing the plugin in `app.component.ts`, here we will set the `GoogleAnalytics` tracking ID and this will get us started:

```
// src/app/app.component.ts
import { GoogleAnalytics } from '@ionic-native/google-analytics';
 constructor(private ga: GoogleAnalytics) { }
 platformReady() {
 // Call any initial plugins when ready
 this.platform.ready().then(() => {
 Splashscreen.hide();

 // Google Analytics
 return GoogleAnalyticsthis.ga.startTrackerWithId("UA-
   92660667-1")
 .then(() => {
 console.log('Google analytics is ready now');
 return
 GoogleAnalyticsthis.ga.enableUncaughtExceptionReportin
  g(true)
 }).then((_success) => {
 console.log("startTrackerWithId success")
 }).catch((_error) => {
 console.log("enableUncaughtExceptionReporting",
    _error)
 })
 });
 }
```

The next step is to integrate the Analytics plugin inside the pages so that we can track the activities:

```
// src/pages/home/home.ts

import { GoogleAnalytics } from '@ionic-native/google-analytics';

constructor(public platform: Platform, private ga:
    GoogleAnalytics) {
 this.platform.ready().then(() => {
 GoogleAnalyticsthis.ga.trackView("Home Page");
 });
}

trackEvent() {
 let active = this.slider.getActiveIndex();
 this.platform.ready().then(() => {
 GoogleAnalyticsthis.ga.trackEvent("Slider", "Slider-
 Changed", "Label", active);
 });
 }
```

We have initialized the `trackView()` method and passed it with the specific page name we are on. Also, we are tracking events if someone manually changes the slide for slider. Similarly, we can track transactions for a purchase inside the application. There are multiple methods available that can be used according to the requirements. For the latest updated methods, check the official documentation for the plugin: `https://ionicframework.com/docs/v2/native/google-analytics/`.

Ionic deeplinks

The Ionic deeplinks plugin handles the custom URL scheme and universal app links for Android and iOS. We need to make sure that we don't confuse this with the deeplinker API that handles registering and displaying views based on URLs. We need to think that navigation is not controlled by URLs; they are mainly used as breadcrumbs. This is mainly useful when we convert the Ionic app to PWA, as web apps require URLs for SEO and direct navigation purposes.

We will use both the deeplink plugin and deeplinked APIs for opening custom URLs directly inside a Native application and web application, respectively. Let's see first how to install the deeplink plugin:

```
$ ionic cordova:plugin add ionic-plugin-deeplinks --
  variable URL_SCHEME=vplanet --variable
  DEEPLINK_SCHEME=https --variable
  DEEPLINK_HOST=vplanetcommerce.com --variable
  ANDROID_PATH_PREFIX=/

$ npm install @ionic-native/deeplinks —save
```

We use both the deeplink plugin and deeplinked APIs inside our application for opening custom URLs directly inside a native application and web application, respectively. Let's see first how to install the `Deeplink` plugin:

```
// src/app/app.component.ts
import { Deeplinks } from '@ionic-native/deeplinks';
constructor(private deeplinks: Deeplinks) { }
this.platform.ready().then(() => {
 Splashscreen.hide();

 // Convenience to route with a given nav
  this.dDeeplinks.routeWithNavController(this.nav, {
  '/about-us': AboutPage,
  '/categories': CategoriesPage,
  '/wishlist': WishlistPage,
  '/cart': ShowcartPage,
  '/login': LoginPage,
  '/settings': SettingsPage,
  '/categories/:categoryId':ProductsPage,
  '/products/:productId': ProductDetailPage
 }).subscribe((match) => {
 // match.$route - the route we matched
 // match.$args - the args passed in the link
 // match.$link - the full link data
 console.log('Successfully routed', match);
```

```
  }, (nomatch) => {
  console.warn('Unmatched Route', nomatch);
  });
  })
```

Let's look at how you can configure `DeeplinkConfig` for web applications. As previously mentioned, these are just used as breadcrumbs, although the same `NavController` instance `push` and `pop` operations are used to navigate inside the application:

```
// src/app/app.module.ts

import { IonicApp, IonicModule, IonicErrorHandler, DeepLinkConfig } from
'ionic-angular';
import { APP_BASE_HREF } from '@angular/common';

imports: [
 IonicModule.forRoot(vPlanetApp, {locationStrategy: 'hash'},
deepLinkConfig),
 Ionic2RatingModule,
 TranslateModule.forRoot({
 provide: TranslateLoader,
 useFactory: createTranslateLoader,
 deps: [Http]
 })
],
providers: [
 { provide: ErrorHandler, useClass: IonicErrorHandler},
 { provide: Storage, useFactory: provideStorage },
 { provide: APP_BASE_HREF, useValue: '/'},
 UserService,
 DataService
]

// Deeplink Configuration
 export const deepLinkConfig: DeepLinkConfig = {
 links: [
 { component: HomePage, name: 'Home Page', segment: '' },
 { component: CategoriesPage, name: 'Categories Page',
   segment: 'categories' },
 { component: ProductsPage, name: 'Categories Product
   Page', segment: 'categories/:categoryId' },
 { component: ProductDetailPage, name: 'Product Details
   Page', segment: 'products/:productId' },
 { component: WishlistPage, name: 'Wishlist Page',
   segment: 'wishlist' },
 { component: ShowcartPage, name: 'Showcart Page',
  segment: 'cart' },
 { component: SupportPage, name: 'Support Page', segment:
```

```
  'feedback' },
{ component: SettingsPage, name: 'About Page', segment:
  'settings' },
{ component: AboutPage, name: 'About Page', segment:
  'about' },
{ component: LoginPage, name: 'Login Page', segment:
  'login' },
{ component: SignupPage, name: 'Signup Page', segment:
  'signup' },
{ component: AccountPage, name: 'Account Page', segment:
  'account' }
]
};
```

Now, similar to how you open a specific link with a unique URL, you can go to a specific view with a custom URL scheme. So now our application can be accessed at `vplanet://wishlist`, which will now open the application's `wishlistPage` directly.

Facebook connect

The Facebook plugin is one of the most loved and widely used plugins for social integrations. With Ionic Native and improved documentation, now it's really easy to integrate the Facebook connect plugin inside your application. The Facebook plugin allows you to use the same JavaScript code in your Cordova application as you use in your web application. The best part about this plugin is it uses the Native Facebook app to perform Single Sign On for the user. If the Facebook Native app is not installed, then the sign on will degrade gracefully using the standard dialog based authentication. This plugin just has support for iOS and Android for now. Before installing this plugin you need to create a Facebook app via `developers.facebook.com` and then copy the `appId` and `appName`, which will be required while adding the plugin:

We are now ready to install the plugin for Facebook, which requires APP_ID and APP_NAME:

```
$ ionic cordova:plugin add cordova-plugin-facebook4 --
variable APP_ID="419376571536241" --variable
APP_NAME="chatApp"
```

Other than this, there are a set of requirement checks that need to be done for specific platforms. For Android you need to configure the project with your fb_ app _id in the res/values/facebookconnect.xml file:

```
<resources>
  <string name="fb_app_id">419376571536241</string>
  <string name="fb_app_name">chatApp</string>
</resources>
```

For iOS, you have to change the following:

- Change FacebookAppID to project *-info.plist

- Change the URL scheme to fb<YOUR APPID>: for example, fb 419376571536241

You are now ready to use Facebook plugin methods that the $codovaFacebook module provides. The following are some of the methods that we have with Ionic Native:

- login(permissions): A string of arrays, such as ['public_profile', 'email']
- showDialog(options): A JSON object with three keys; method, link, and caption
- api(path, permissions): A Facebook API path query, for example, api("me", ["public_profile"])
- getLoginStatus(): Checks if a user is logged in or not
- getAccessToken(): Fetches the access token of the current logged in session
- logout(): Logs out from Facebook

For testing in a browser, you still need to make some changes in the application config:

```
let appID = 419376571536241;
let version = "v2.0"; // or leave blank and default
is v2.0

constructor() {
    Facebook.browserInit(appID, version);
```

```
}
```

Also, you have to set the redirect URL, which will look as follows. As Ionic servers run on port 8100, you can give `http://localhost:8100` as a redirect URL:

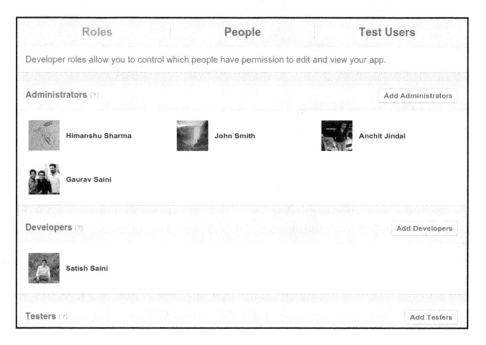

Building a music player

We will now build a media player application, where you will have audio and video streaming and download options to save the file to your mobile. We will be mainly integrating a media plugin for audio streaming via `ionic-audio` package, a `StreamingMedia` plugin will be used for playing the videos, and a `FileTransfer` plugin for downloading the files to a local system directory.

Let's first create an application for which we will be using the tabs template:

```
gaurav@gaurav-thinkpad:~/projects$ ionic start jingle-app tabs --v2
Creating an Ionic 2.x app in /home/gaurav/projects/jingle-app based on the tabs template.

Downloading: https://github.com/driftyco/ionic2-app-base/archive/master.zip
Downloading: https://github.com/driftyco/ionic2-starter-tabs/archive/master.zip
Installing npm packages (may take a minute or two)...
/
♫ ♪ ♫ ♪  Your Ionic app is ready to go! ♫ ♪ ♫ ♪

Some helpful tips:

Run your app in the browser (great for initial development):
  ionic serve

Run on a device or simulator:
  ionic run ios[android,browser]

Share your app with testers, and test on device easily with the Ionic View companion app:
  http://view.ionic.io
gaurav@gaurav-thinkpad:~/projects$ ▊
```

We just now need to add the **ionic platform add android** platform and we will be ready to integrate the plugins and build our jingle application.

Media, streaming, and transfer plugin

We will be starting with a simple music player application that will demonstrate the usage of different plugins, but the possibilities here are endless. You might think of building a music recommendation application that will play music according to your inputs, mood, or other factors, although you will need a good backend application ready for this to support recommendations.

As we have decided on the major functionality that our application will have, we will now need mainly three Cordova plugins, which we will install now:

```
$ ionic cordova:plugin add cordova-plugin-media
$ ionic cordova:plugin add cordova-plugin-file-
  transfer
$ ionic cordova:plugin add cordova-plugin-streaming-
  media
```

The next step will be to install their respective NPM packages and ionic-audio module, which we will use for the audio player:

```
$ npm install @ionic-native/media @ionic-
  native/filetransfer @ionic-native/streaming-media
  ionic-audio --save
```

Let's integrate the ionic-audio module, this module automatically switches between `WebAudioProvider` and `CordovaMediaProvider` according to the environment, which is really helpful while testing the application, as we can easily verify the application inside the browser itself:

```
// src/app/app.module.ts
import { IonicAudioModule, AudioProvider, WebAudioProvider,
audioProviderFactory } from 'ionic-audio';
import { NgModule, ErrorHandler, CUSTOM_ELEMENTS_SCHEMA } from
'@angular/core';
import { IonicApp, IonicErrorHandler, IonicModule }
        from 'ionic-angular';

@NgModule({
 imports: [
 IonicModule.forRoot(MyApp),
 IonicAudioModule.forRoot({ provide: AudioProvider,
 useFactory: audioProviderFactory })
 ],
 schemas: [ CUSTOM_ELEMENTS_SCHEMA ]
})
```

We have three tab views, `HomePage`, `MusicPage`, and `VideoPage`. Audio Player is integrated on the `MusicPlayerPage`. We will inject `AudioProvider` for controlling the audio player and its functions:

```
// app/pages/music-player/music-player

import { AudioProvider } from 'ionic-audio';

export class MusicPlayerPage {

 source: string;
 img: string;
 singleTrack: any;
 allTracks: any[];
 selectedTrack: number;

 constructor(
 public platform: Platform,
 public navParams: NavParams,
 public viewCtrl: ViewController,
 private _audioProvider: AudioProvider) {

 this.source = this.navParams.get('url');
 this.img = this.navParams.get('img');
```

```
this.singleTrack = {
src: this.source,
artist: this.source.slice(25,-4),
title: this.source.slice(25,-4),
art: this.img,
preload: 'metadata' // tell the plugin to preload
metadata such as duration for this track, set to 'none' to turn
off
};
}

ngAfterContentInit() {
// get all tracks managed by AudioProvider so we can
   control playback via the API
this.allTracks = this._audioProvider.tracks;
}

playSelectedTrack() {
// use AudioProvider to control selected track
this._audioProvider.play(this.selectedTrack);
}

pauseSelectedTrack() {
// use AudioProvider to control selected track
this._audioProvider.pause(this.selectedTrack);
}

onTrackFinished(track: any) {
console.log('Track finished', track)
}

}

// src/pages/music-player/music-player.html

<ion-content padding>
 <ion-card>
 <img [src]="img"/>
 <ion-card-content>
 <ion-card-title>
 {{ source | trimurl}}
 </ion-card-title>

 <audio-track #track [track]="singleTrack">
 <ion-item [audioTrack]="track">
 <ion-thumbnail item-left>
 <img src="{{track.art}}">
 <audio-track-play light [audioTrack]="track">
```

```
    <ion-spinner></ion-spinner>
    </audio-track-play>
    </ion-thumbnail>

    <div item-content style="width:100%">
    <p><strong>{{track.title}}</strong> ○ <em>
    {{track.artist}}</em></p>
    <audio-track-progress-bar duration progress
    [audioTrack]="track"></audio-track-progress-bar>
    <em style="font-size:.5em">Track ID: {{track.id}}</em>
    </div>
    </ion-item>
    </audio-track>
    </ion-card-content>
    </ion-card>
</ion-content>
```

We have added a respective component for the ionic-audio plugin. Next we will be able to download the songs via the FileTransfer plugin:

```
// src/pages/music/music.ts

import { Transfer, TransferObject } from '@ionic-
native/filetransfer';
```

```
declare var cordova: any;

@Component({
 selector: 'page-music',
 templateUrl: 'music.html'
})

export class MusicPage {

 constructor(private transfer: Transfer);


 download(song) {
 const fileTransfer = new Transfer();
 const fileTransfer: TransferObject =
 this.transfer.create();
 let url = song.url;

 fileTransfer.download(url,
 cordova.file.externalRootDirectory +
 song.name).then((entry) => {
 alert('download complete: ' + entry.toURL());
 }, (error) => {
 // handle error
 });
 }
}
```

Make sure you declare the cordova variable before the @Component decorator. Also, we need to create an instance of the Transfer() method and then pass the URL and path to it where it will save the file. We have set the rootDirectory currently where the files will save.

Next, we will be using the Streaming plugin for playing videos inside our application. The streaming plugin uses the Native media player, which is really smooth, rather than playing inside HTMLAudioElement:

```
// src/pages/video-player/video-player
import { Component } from '@angular/core';
import { NavController, NavParams } from 'ionic-angular';
import { StreamingMedia, StreamingVideoOptions } from '@ionic-
native/streaming-media';

@Component({
 selector: 'page-video-player',
 templateUrl: 'video-player.html'
})
```

```
export class VideoPlayerPage {

    source: string;

    constructor(public navCtrl: NavController, public
    navParams: NavParams) {

    this.source = this.navParams.get('url');

    // Playing a video.
    let options: StreamingVideoOptions = {
    successCallback: () => { console.log('Video played')
    },
    errorCallback: (e) => { console.log('Error streaming')
    },
    orientation: 'landscape'
    };

    StreamingMedia.playVideo(this.source, options);
    }

    ionViewDidLoad() {
    console.log('ionViewDidLoad VideoPlayerPage');
    }

}
```

We can further extend the music player with more features such as adding songs as favorites, creating playlists, or even adding a music control plugin, which will display media notifications with play/pause, previous, and next actions. There are unlimited possibilities available as Cordova plugins are evolving quickly.

Until now, we have mostly focused on functions that the Ionic Native wrapper provides rather than an official plugin. The reason for this is to take advantage of the ease of development and integration with your existing Ionic applications. I recommend everyone to check out the official plugin documentations to see any new improvements and changes. Also, sometimes there might be a change in official repository location, so this way we will always be updated. Hybrid development is going at a good pace after the launch of the Ionic Framework, the same goes for the Cordova plugin development, so it's always good to verify with official plugin documentation for new features.

Cordova versus phonegap

There has been a lot of confusion and questions about Cordova and phonegap. Also, many people ask questions about which to use with Ionic. Although Ionic uses Cordova, we can also definitely use phonegap. Many still don't know about Cordova as phonegap is before phonegap.

Let's quickly look back at its history; phonegap was created in 2009 by Nitobi. Later in 2011, Adobe took over Nitobi along with the rights to the phonegap platform. The Open source core code of the platform was donated to the Apache organization for further development. Phonegap is also free and open source. Initially, there was minimal difference between Cordova and phonegap, but slowly Adobe started building a proprietary set of services around phonegap. Phonegap build is one of those services.

Cordova is the engine that powers phonegap, similar to webkit, which is the engine for the Chrome browser. There have been long debates upon which platform to use, there have been conflicts of interest between developers. Many criticize phonegap as a patch up version of Cordova, only using phonegap for its features such as the remote build option. At the end this is nothing to do with Ionic and it's independent of this debate. As a developer, you are free to use any of these.

If you still want to use phonegap and take advantage of the services built, you can easily choose that while building the project with phonegap. Ionic CLI commands use Cordova, so you will need to install and use Cordova.

Summary

There will be a continuous evolution of Cordova plugins and we need to take care that we are updated with the latest features that are introduced. With Ionic coming into the market, the last few years have seen increased development of hybrid applications and plugins. This is because more people are interested in using it and multiple custom plugins are developed for keeping in touch with the latest technology. One similar example is Bluetooth beacons and multiple plugins developed by developers for supporting its different formats. We have now covered plugin development, which is another important aspect of developing complex hybrid applications. We can now leverage the power of plugins for building advance hybrid applications that require access to Native APIs. Later in the chapter, we went on to cover many Ionic CLI commands that help us quickly develop and debug applications. Ionic CLI is advancing day by day and many new features and tasks are coming. What we all should do is regularly update Ionic CLI and check for new features so that in our next applications we can take advantage of that.

In the Chapter 4, *Ionic Platform and Services*, we will look at Ionic services and tools. Over the last two years, Ionic has evolved a lot and the Ionic team has built really useful tools that speed up development time. Ionic CLI is one of the most used and important tools. Ionic as a platform offers Ionic push, Ionic deploy, Ionic analytics, and Ionic lab, along with creator and playground help to build prototypes quickly. The Ionic team is working really hard to bring us all this awesome stuff that helps us to make innovative applications for users.

4
Ionic Platform and Services

So far, we have looked at almost all the key components for building an enterprise level application. Ionic platform is a set of beautiful tools and services that speed up the process of development and streamline everything on a single platform from building, testing, debugging, and deploying applications. With the Ionic team continuously working on improving Ionic platform and tools, you will see a lot more in the near future. We will cover almost all services and tools that Ionic provides so you can utilize your project:

- Ionic Cloud
- Ionic Auth
- Ionic DB
- Ionic push
- Ionic deploy
- Ionic package
- Ionic View, Creator, and Playground

Ionic Cloud

Ionic started as a framework called Ionic framework, which has now become a complete platform with the launch of `Ionic.io` and tools and services involved with it. `Ionic.io` or Ionic Cloud Platform provides rich and hybrid-focused backend services and tools that ease the process of development and help build enterprise applications at a rapid pace. Ionic platform at present is out from beta phase and they have launched services now on some paid subscription models as it's necessary to sustain and generate revenue for the Ionic team. But the Ionic framework will always be open source; just the backend services will have some pricing plans.

Getting started with the Ionic platform is as simple as signing up for an account. After this, you will be able to share, upload, and use all other services provided by the platform.

Installing Ionic Cloud

Setting up Ionic Cloud is really simple; we need to install the Ionic Cloud npm package and hook Ionic account, which is just required for using Ionic services. We will be integrating cloud services to our existing vPlanet-commerce application. Recently, Ionic CLI v3 has launched and is currently in beta and it have changed a lot many things and changed the commands:

```
$ npm install @ionic/cloud-angular --save
$ ionic login
$ ionic link
```

> Be sure not to confuse Ionic CLI v3 and Ionic v3, which are completely different. Ionic CLI contains primary tools for the development of Ionic apps.

To update an Ionic 2 or 3 based project to work with Ionic CLI v3, we have to run the following command.

```
$ npm install --save-dev @ionic/cli-build-ionic-angular@beta @ionic/cli-plugin-cordova@beta
```

Now, when you will link the project to an existing project or create a new project, make sure you have logged in to Ionic Cloud which will ask for the username and password of the Ionic account, which you might have already created at ionic.io. After this you can upload your snapshot to Ionic Cloud.

```
gaurav@gaurav-thinkpad:~/projects/vplanet-commerce$ /usr/bin/ionic login
Log into your Ionic account
If you don't have one yet, create yours by running: ionic signup

? Email: ga          @gmail.com
? Password: ************
[OK] You are logged in!
gaurav@gaurav-thinkpad:~/projects/vplanet-commerce$ /usr/bin/ionic link
? Which app would you like to link ionic2-conf (acb64cbd)
[OK] Project linked with app acb64cbd!
gaurav@gaurav-thinkpad:~/projects/vplanet-commerce$ /usr/bin/ionic upload
✓ Requesting snapshot - done!
[===============] 100% 0.0s
✓ Uploading snapshot - done!
[OK] Uploaded snapshot c4fdba46-a370-4b51-89cf-fa22c9ccd0a0!
```

The next step is to import `CloudSettings` inside our `app.module.ts`:

```
// src/app/app.module.ts


import { CloudSettings, CloudModule } from '@ionic/cloud-angular';

const cloudSettings: CloudSettings = {
  'core': {
    'app_id': 'f8fec798'
  }
};

@NgModule({
  declarations: [ ... ],
  imports: [
    IonicModule.forRoot(vPlanetApp),
    Ionic2RatingModule,
    TranslateModule.forRoot({
      provide: TranslateLoader,
      useFactory: createTranslateLoader,
      deps: [Http]
    }),
    CloudModule.forRoot(cloudSettings)
  ],
  bootstrap: [IonicApp],
  entryComponents: [ ... ],
  providers: [ ... ]
})
```

We have successfully now uploaded our vPlanet-Commerce application to `ionic.io`. We are now ready to start using Ionic services such as Push, Deploy, DB, and so on. You can find all your applications on the Dashboard (`https://apps.ionic.io/apps`):

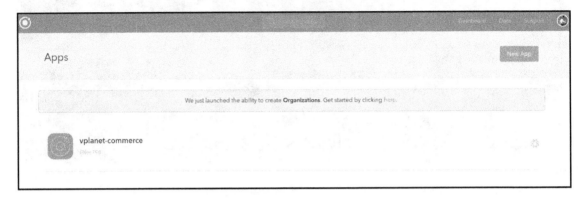

You can find all the information about the application inside the **Settings** tabs. After you click on your specific project, you will be able to see details about specific projects. Currently, we are using a free plan for our vPlanet-Commerce application, which gives around 5000 deploys, 10,000 push notifications, and 100 package builds. You can upgrade the plan at anytime as you will see that your application users are increasing and traffic is growing. What's best is that we will be integrating each and every Ionic Cloud service starting from Auth for user login and sign-up process. Ionic DB as our application database, Push Notification, Package, and Deploy:

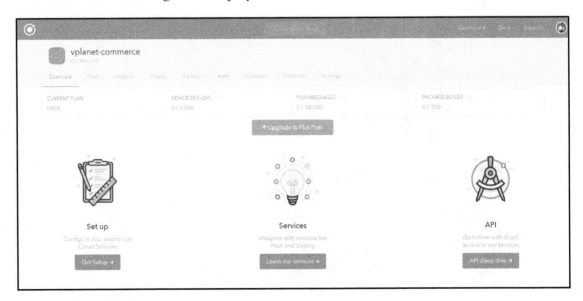

Also, another important point to note is that when you have completed setting up the Ionic Cloud, you will find `app_id` and `api_key` inside `.io-config.json` and some other app details inside `ionic.config.json`:

```
// .io-config.json
{"app_id":"f8fec798","api_key":"5b22exxxxxxxxxxxxxxxxxxxxxxxxxxxxxxxxxxxxxxxx
xe68e"}

// ionic.config.json

{
  "name": "vplanet-commerce",
  "app_id": "f8fec798",
  "v2": true,
  "typescript": true
}
```

After this chapter completion, you will be almost in mid of the book and in half way itself you are now able to build and application with Ionic, Integrate various Ionic components, Cordova plugins and able to integrate with Ionic Cloud services and finally deploy the application to the play store.

Ionic Auth

User authentication is an essential component of most apps. It allows you to identify your users and provide a unique experience for them. Logging in/out, managing user data, and other API features for auth are handled by authentication providers. The setup and usage may vary per provider. There are multiple providers available with Ionic Auth:

- E-mail/password
- Custom
- Facebook
- Google
- Twitter
- Instagram
- LinkedIn
- GitHub

We will be integrating e-mail/password provider in our vPlanet application. We will also be integrating Facebook login in Chapter 6, *TasteBite App with Firebase,* where we will be building our tastebit application with Firebase. e-mail/password authentication is really easy to configure inside our application. We will start with integrating a login method inside our `LoginPage`, where first we will import `Auth` from the `@ionic/cloud-angular` package:

```
// src/pages/login/login.ts

import { Auth, IDetailedError} from '@ionic/cloud-angular';

export class LoginPage {
  login: {email?: string, password?: string} = {};
  submitted = false;

  constructor(public navCtrl: NavController,
    public loadingCtrl: LoadingController,
    public auth: Auth,
    public toastCtrl: ToastController,
    public events: Events) { }

  onLogin(form: NgForm) {
    this.submitted = true;

    if (form.valid) {

      // start Loader
      let loading = this.loadingCtrl.create({
        content: "Login wait...",
        duration: 20
      });
      loading.present();

      this.auth.login('basic', this.login).then((result)
      => {
        // user is now registered
        this.navCtrl.setRoot(HomePage);
        this.events.publish('user:login');
        loading.dismiss();
        this.showToast(undefined);
      }, (err: IDetailedError<string[]>) => {
        console.log(err);
        loading.dismiss();
        this.showToast(err)
      });
    }
  }
```

```
showToast(response_message:any) {
  let toast = this.toastCtrl.create({
    message: (response_message ? response_message : "Log In
Successfully"),
    duration: 1500
  });
  toast.present();
}

onSignup() {
  this.navCtrl.push(SignupPage);
}
}
```

We have integrated this with a login form that gets the data as an object and we pass it in `this.auth.login()` method, as you can see in the preceding code. This will return token and various other data that is stored inside local storage by default. In the following screenshot you can see mainly three key/value pairs that are saved, `ionic_auth`, `ionic_insights_session`, and `ionic_user`:

Next, we integrated this with a login form that gets the data as an object and we pass it in `this.auth.login()` method, as you can see in the preceding code. This will return token and various other data that is stored inside Local Storage by default. In the preceding screenshot, you can see mainly three key/value pairs. User data is mainly stored inside `ionic_user`, where you will find username, e-mail, and other details.

Let's now look into registering a new user inside our application. We will be integrating it within our `SignupPage`:

```
// src/pages/signup/signup.ts

import { Component } from '@angular/core';
import { NgForm } from '@angular/forms';
import { NavController } from 'ionic-angular';
import { LoginPage } from '../login/login';
import { Auth, UserDetails, IDetailedError } from '@ionic/cloud-angular';
```

```
@Component({
  selector: 'page-user',
  templateUrl: 'signup.html'
})
export class SignupPage {

  signup: {name?: string, email?: string, password?: string} = {};

  details: UserDetails = {
    "name": this.signup.name,
    "email": this.signup.email,
    "password": this.signup.password
  };

  submitted = false;

  constructor(
    public navCtrl: NavController,
    public userService: UserService,
    public auth: Auth) {}

  onSignup(form: NgForm) {
    this.submitted = true;

    if (form.valid) {
    // registering new user
      this.auth.signup(this.signup).then(() => {
        // user is now registered
        this.navCtrl.push(LoginPage);
      }, (err: IDetailedError<string[]>) => {
        for (let e of err.details) {
          if (e === 'conflict_email') {
            alert('Email already exists.');
          } else {
            // handle other errors
          }
        }
      });
    }
  }
}
```

All the common user data that we pass during the signup process is saved in the `details` attribute. Data such as username, e-mail, password, name, and image URL can be passed as an object to `this.auth.signup()` method. Other than this, if we need to store any custom data we have this ability and we can set, get, or unset custom data.

We have one such example on our `AccountPage` where we have an option to Update the Date of Birth:

```
// src/pages/account/account.ts

import { User } from '@ionic/cloud-angular';

constructor(
    public alertCtrl: AlertController,
    public nav: NavController,
    public user: User) {
  }

export class AccountPage {
  birthdate: string;

  constructor(
    public alertCtrl: AlertController,
    public nav: NavController,
    public user: User) {
  }

  // Present an alert for adding date of birth
  // clicking OK will update the DOB
  // clicking Cancel will close the alert and do nothing
  addDOB() {
    let alert = this.alertCtrl.create({
      title: 'Add Date of Birth',
      buttons: [
        'Cancel'
      ]
    });
    alert.addInput({
      name: 'birthdate',
      value: this.birthdate,
      placeholder: 'birthdate'
    });
    alert.addButton({
      text: 'Ok',
      handler: (data: any) => {
        this.user.set('birthdate', data.birthdate);
        this.user.save();
      }
    });

    alert.present();
  }
```

```
logout() {
  this.userService.logout();
  this.nav.setRoot(LoginPage);
}

}
```

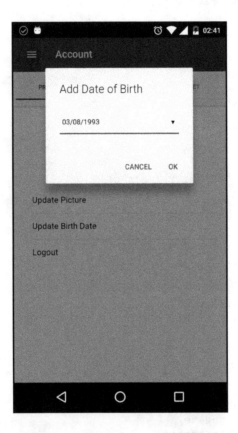

All the changes you do when you set custom data will be in your local changes, unless you called the `save()` method, which will update it to the server. You can implement an update profile page where you can get custom data from the user and then when clicking the **Save** button you can save it to the server.

We have logout, `auth` status, and password reset functions available within Ionic Auth service. In the `logout()` method it will clear the current user context, while `isAuthenticated()` will return the boolean value whether the user is authenticated or not:

```
// src/providers/user-service.ts

logout() {
    this.auth.logout();
    this.push.unregister();
    this.events.publish('user:logout');
    return true;
}
```

For password reset there are two options, using the client reset form or using the hosted reset form. I personally prefer building reset functionality within an application using the client method. We can just ask the user to enter their e-mail address in an alert and then show another alert with two fields for the six digit reset code you will receive on your `email` and `newPassword`:

```
// for submitting the email
this.auth.requestPasswordReset(email);

// for updating new password
this.auth.confirmPasswordReset(resetCode, newPassword);
```

Now we have integrated most of the Ionic Auth features in our application and have a good overview of how these services work. Just to make sure you are updated with any new features coming in the future, do check the Ionic official documentation (`https://docs.ion ic.io/services/auth/`).

Ionic DB

Ionic DB is the most recent addition in Ionic Cloud services. IonicDB is used to securely store your application data remotely. It integrates well with Auth so users can securely store their data and retrieve it in real time. We will be fetching data in our application from IonicDB and for that we all set to integrate it directly as we have already installed and configured cloud client. Some important features of IonicDB are:

- JSON style document storage
- Real-time data updation
- Authentication and permission manages the read/write privileges.

We will start with creating a database for our application in the dashboard and adding specific collections to it. Collections are a group of documents that are identified by a unique key stored in the id field. You can also think collections similar to what we have tables in relational database. We will be creating some initial collections categories, parent_categories, popular_products, and products, which we will be using inside our application in different views. You will also find user collection, which is automatically created as we have integrated Auth in our application:

We will be using Ionic DB inside multiple views in our application, so for this we have created a DataService provider, which will be directly communicating with Ionic DB and returning data to the views from where the service will be called. This will help us save a lot of code duplicity. Before we get started let's first check out the process of how the Ionic DB works and fetches the data:

```
// src/providers/data-service.ts

import { Injectable } from '@angular/core';
import { Storage } from '@ionic/storage';
import { Database } from '@ionic/cloud-angular';
import { Observable } from 'rxjs/Observable';

import 'rxjs/add/operator/map';
import 'rxjs/add/operator/catch'

@Injectable()
export class DataService {
```

```
cartItems: any[];
loading: any;

constructor(public db: Database) {
  // Establish Initial connection
  this.db.connect();
}

public getCategories(): Observable<any> {
  const categories:any = this.db.aggregate({
    name: "testing",
    categories: this.db.collection('categories'),
    parent_categories:
    this.db.collection('parent_categories')
  })
  return categories.fetch();
}

public getPopularProducts(): Observable<any> {
  const popular_products:any =
  this.db.collection('popular_products');
  return popular_products.order("price",
  "ascending").fetch();
}

}
```

Before fetching data from IonicDB collections we need to establish an initial connection with the database, after that, we will be able to create collections. You can see in the preceding code that we have established a connection in our constructor method and after that we have declared multiple functions for fetching data from the database.

There are multiple read and write operations available for collections. Inside our `getPopularProducts()` function, we have the `order()` method, which is retrieving products that are according to low to high price.

There can be a scenario where we need to query multiple collections and combine the result into a single JSON response. On `CategoriesPage` we are doing the same fetching categories and `parent_categories` at the same time. So we can prepare a tree to display it accordingly on the UI side. The `aggregate()` method combines the results of multiple DB queries into one result.

You may have noticed that we have used the `fetch()` method here rather than the `watch()` method, which we used for the `getPopularProducts()` function. The difference between `watch()` and `fetch()` is that `watch()` updates the data in real time:

```
// src/pages/home/home.ts

import { DataService } from '../../providers/data-service';

export class HomePage {
  constructor(public data: DataService) {
    this.data.getPopularProducts().subscribe((popular) => {
      this.products = this.arrayToMatrix(popular, 2);
      console.log(this.products)
    }, (error) => {
      console.error(error);
    });
  }
}

// src/pages/categories/categories.ts

this.data.getCategories().subscribe((data) => {
  for (var i = 0; i < data.categories.length; ++i) {
    let p_id = data.categories[i].parent_id;
    for (var j = 0; j < data.parent_categories.length;
    ++j) {
      if(data.parent_categories[j].id == p_id)
      {
        data.parent_categories[j].child.push(data.categories[i])
      }
    }
  }
  this.categoryList = data.parent_categories;
}, (error) => {
  console.error(error);
});
```

All the read and write operations return the RxJS observable. `subscribe` is a method on observable, which takes three functions as parameters:

- `next(result)`: callback method that receives collection data
- `error(error)`: callback method that receives error data
- `complete()`: callback executed when the result set is iterated completely

If you want to stop receiving emissions you can call the unsubscribe() method. We will be mostly using read methods from our mobile application to retrieve data from the database. Although, we also have multiple write operations that can also be used for user centric applications such as to-do applications.

Let's see some write operations also and how we can use it. Although, in our vPlanet application we haven't used write operations, but let's demonstrate how we can use some write operations for a to-do application:

```
const todoItems = this.db.collection("items");

// Insert a todo item document

let dateTime = new Date().toLocaleString();

todoItems.insert({
    id: "1",
    user: this.user.id,
    text: "Prepare project presentation",
    status: "pending"
    timestamp: dateTime
});

// Update todo item

let dateTime = new Date().toLocaleString();

todoItems.update({
    id: "1",
    user: this.user.id,
    text: "Prepare project presentation",
    status: "completed"
    timestamp: dateTime
});
```

We now have a good idea of how read and write operations work with IonicDB. IonicDB is still in active development, so you can expect some issues, but until now I have had a smooth journey developing vPlanet applications.

Authentication and permissions

Ionic DB integrates existing auth services, which is disabled by default, as in our vPlanet application we don't need authentication to be enabled as most of the data should be publicly available. In case of to-do applications we need to enable authentication in our IonicDB, which will require users to log in using Ionic Auth before they can connect to the database. This will allow users to access data only after they are authenticated. For enabling authentication we need to update some settings in `app.module.ts`:

```
// src/app/app.component.ts

const cloudSettings: CloudSettings = {
  'core': {
    'app_id': 'APP_ID'
  },
  'database': {
    'authType': 'authenticated'
  }
};
```

Also, another important point is to make sure that you establish database connections only after the login method is called or the user is already authenticated:

```
if(this.auth.isAuthenticated()){
  // Connect DB aftere checking user authentication
  this.db.connect();
} else {
  // Need to login
  this.auth.login('basic', details).then( () => {
    // Now it is safe to connect
    this.db.connect();
  });
}
```

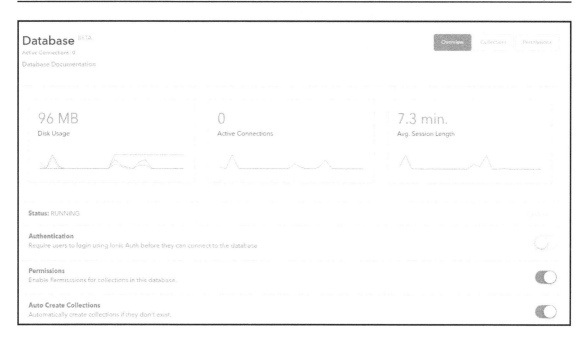

Permissions allow you to restrict access to the database collections. We can control what data users can read and write. Permissions are based on white-listing, when it is enabled from the dashboard we will not be able to access data by default. We need to create permission for each collection and define them as default or authenticated groups. Default group permissions are applied to all users, whether they are authenticated or not. Authenticated group permissions are applied to authenticated users only.

Let's get back to our todo app demonstration, now here we have another concern that only data that is related to users should be accessed by that user. To-do items created by one user login should not have permission to other users.

So basically it should be filtered according to current users. For this we use custom permissions, for example, while inserting an item we specified the `userId` to which the item is associated:

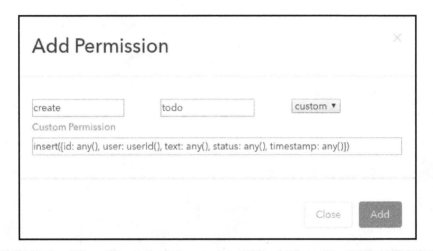

So now while fetching the items we will pass the current `userId`, which will retrieve data in the current user context:

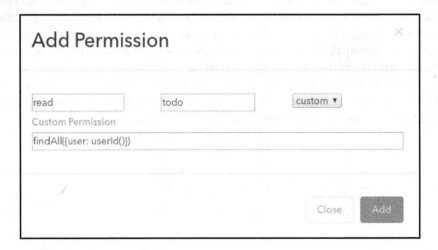

```
todoItems.findAll({user:this.user.id}).watch().subscribe( (items) => {
    // this will only fetch items that
    // login user have created
    console.log(items);
})
```

We have tried here to touch each possible section available with IonicDB and get with some sample code. This will really ease the initial learning curve and get you directly started on building the applications using IonicDB. With IonicDB we can also integrate that in our backend NodeJS application.

Ionic push

Ionic push allows you to send messages and alerts to a user's device. Push notifications can be triggered even when application is closed. Push notifications have now been around for a long time and are commonly used in every application. Many of you probably already have good experience with integrating GCM and APNS in the past. So here our main interest is how we can use these services from the Ionic push service. Ionic push allows you to send notifications to both platforms from a single interface or API. With the Ionic push service we can send notifications to:

- Target users
- Schedule notifications
- Send contextual content
- Attach media to notifications
- Automatically send notifications on custom logic
- Trigger app events with background notifications

Before we can send notifications from the Ionic push service we need to set up FCM Project and API keys. We can find the FCM server key and sender ID in the firebase console settings and cloud messaging tab:

- The next step is to generate the `keystore` file, which we will be uploading with other details while creating the security profile for our application. While creating keystore it will first ask for the **Keystore Password** and after entering other details it will ask for the **Key Password** for alias. Just make sure you remember this as that will be needed while creating security profiles:

```
$ keytool -genkey -v -keystore vplanet.keystore -alias
vplanet-alias -keyalg RSA -keysize 2048 -validity
10000
```

- Now we have everything ready to create a security profile. The **FCM server Key** from the Firebase console, attach the generated `vplanet.keystore` file, alias, and their respective passwords:

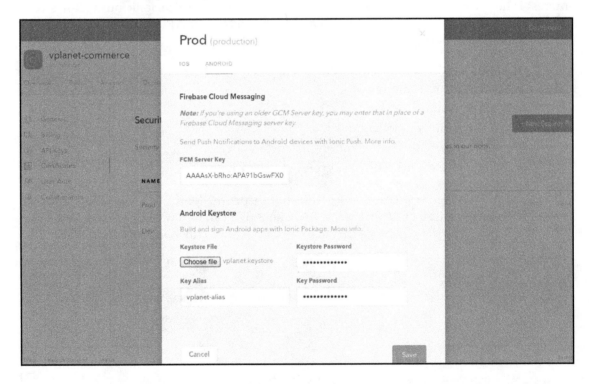

- We will be using phonegap's push plugin to receive `push` notifications. While installing the plugin you need to pass `SENDER_ID`, which you will get from the same place in the Firebase console settings:

```
// Install Cordova Push Plugin
$ionic cordova:plugin add phonegap-plugin-push@1.9.4 -variable
```

```
    SENDER_ID=76xxxxxxxxx --save

// NPM package
$ npm install @ionic-native/push --save
```

We now need to update our cloud configuration in our code in `app.module.ts`:

```
// src/app/app.module.ts

const cloudSettings: CloudSettings = {
  'core': {
    'app_id': 'f8fec798'
  },
  'push': {
    'sender_id': '76xxxxxxxxx',
    'pluginConfig': {
      'ios': {
        'badge': true,
        'sound': true
      },
      'android': {
        'iconColor': '#343434'
      }
    }
  }
};
```

We now have to inject the `push` and integrate it in our application. We will call `Push.init()`, which will create an instance of the `Push` class. We will use this `push` instance to listen to `registration` and `notification` events:

```
// src/app/app.component.ts

import { Push } from '@ionic-native/push';

platformReady() {
    // Call any initial plugins when ready
    this.platform.ready().then(() => {
      this.initPushNotification();
    })
}

initPushNotification() {
    console.log("calling push");
    let push = Push.init({
      android: {
        senderID: "762350093850"
```

```
    },
    ios: {
      alert: "true",
      badge: false,
      sound: "true"
    },
    windows: {}
});

push.on('registration', (data) => {
  console.log("device token ->", data.registrationId);
  //TODO - send device token to server
});

let self = this;

push.on('notification', function(data) {
  TextToSpeech.speak({text: data.message, locale: 'en-
 IN', rate: 0.75})
      .then(() => console.log('Success'))
      .catch((reason: any) => console.log(reason));

  if (data.additionalData.foreground) {
    // if application open, show alert
    alert(data.message);
  }
  //if user NOT using app and push notification comes
  // for demo purpose for every push message we push
  to AboutPage
  // Although you can applu logic and according to API
  data
  // you can push to specific page
  self.nav.push(AboutPage, {message: data.message});

  push.finish(function() {
      console.log("processing of push data is
    finished");
  }, function() {
      console.log("something went wrong with
  push.finish for ID = " + data.additionalData.notId)
  }, data.additionalData.notId);
  });
}
```

Now we need to register the device before we can send `push` notifications. As we are using Ionic auth it's recommended to call the `register()` method after login so that the `Push` token saves to the authenticated user:

```
// src/pages/login/login.ts

this.auth.login('basic', this.login).then((result) => {
  // user is now registered
  // Push Notification register
  this.push.register().then((data: PushToken) => {
    return this.push.saveToken(data);
  }).then((data: PushToken) => {
    console.log('Token saved:', data.token);
  });
}, (err: IDetailedError<string[]>) => {
  console.log(err);
  loading.dismiss();
  this.showToast(err)
});
```

Similarly, we will unregister a device when a user logs out:

```
logout() {
    this.auth.logout();
    this.push.unregister();
    this.events.publish('user:logout');
    return true;
}
```

Let's send a `push` notification from the dashboard; there are multiple options available when you send a message. There are three steps and multiple options available when you send a notification:

- Campaign Name, Subject for Notification, message, payload (key/value pairs), and device options
- Targeting specific user segments or all users

- Scheduling push to send now, later, or based on some trigger. Also, which profile to send notification

The next step from here can be integrating Push APIs in our backend codebase if required. This way there can be multiple possibilities and conditions we can send notifications. Just make sure that you don't send too many notifications to the user's device as many times this is the reason why users uninstall applications.

Ionic deploy

The Ionic deploy service gives you access to live deployments, which helps you publish web assets such as HTML, CSS, and JS directly to your users without uploading to the Play Store every time.

You will have many advantages of using this approach for live deployments:

- Real-time Updates on demand
- Save app stores approval time
- A/B testing by uploading different snapshots to different channels

Before we get started, let's install the deploy plugin, which will update your application on Android and iOS devices:

```
$ ionic cordova:plugin add ionic-plugin-deploy -save
```

We will be injecting the `Deploy` method inside our `app.component.ts`, so every time when the application is opened it checks for new updates:

```
// src/app/app.component.ts

import { AlertController } from 'ionic-angular';
import { Deploy } from '@ionic/cloud-angular';

constructor(
    public deploy: Deploy,
    public alertCtrl: AlertController
  ) {

  // checks if new snapshot available
  this.deploy.check().then((snapshotAvailable: boolean) => {
    if (snapshotAvailable) {
      let alert = this.alertCtrl.create({
        title: 'Update Available !',
        message: 'Do you want to update the application
        now ?',
        buttons: [
          {
            text: 'Later',
            role: 'cancel',
            handler: () => {
              console.log('Cancel clicked');
            }
          },
          {
            text: 'Update',
            handler: () => {
              this.deploy.download().then(() => {
                this.deploy.extract()
              }).then(() => {
                this.deploy.load()
              });
```

```
                }
            }
        ]
    });
    alert.present();
  }
});
}
```

The `check()` method is used to check if an update is available. In case updates are available we present an alert dialog asking the user if they want to update the application now or later.

The next step will be to deploy our snapshot with NOTE and CHANNEL_TAG. There are mainly three channels available for deploy, which are `production`, `staging`, and dev. We can create more channels if required and by default the deploy service will check for production snapshots. If we need to change this then we need to set the `channel` attribute as follows:

```
this.deploy.channel = 'staging';
```

If an invalid channel is found, it will revert to the `production` channel. Also, make sure you update the `channel` attribute before you call the `check()` method:

```
gaurav@gaurav-thinkpad:~/projects/ionic/vplanet-commerce$ ionic upload --note "Initial Deploy" --deploy production
Uploading app...

Saved app_id, writing to ionic.io.bundle.min.js...
Successfully uploaded (f8fec798)

Share your beautiful app with a client or co-worker to test in Ionic View (http://view.ionic.io):

$ ionic share EMAIL

Saved api_key, writing to ionic.io.bundle.min.js...
Deploying to channel: production
Deploy Successful!
gaurav@gaurav-thinkpad:~/projects/ionic/vplanet-commerce$
```

It is worth mentioning that whenever we package code then code is also uploaded to the server. But for deploy to work we need to pass the channel tag:

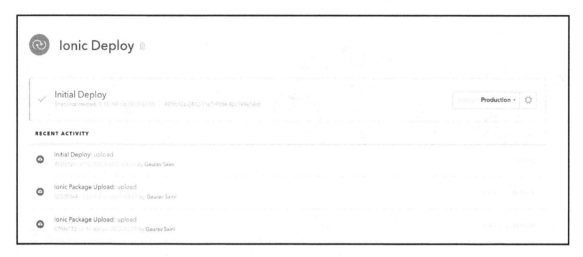

We can test deploy by changing some small HTML changes and then uploading it to the server. Just make sure that you run Ionic server so that the `src` folder is transpired and copied to the `www` folder.

Ionic package

Ionic package is another Cloud service that eases the native build process for different platforms. Basically, `ionic package` builds the Cordova application in the cloud. It is a perfect choice for someone who needs to build iOS apps, but works on Windows or Linux. Some advantages of using the `ionic package` service are:

- Sends builds to others
- Builds apps for platform not supported
- Generates `.apk` and `.ipa` files, which can be directly uploaded to app stores

As we have already set up the security profile and credentials, we can straightaway get started. Here we build our application for `PLATFORM_TAG` Android and `PROFILE_TAG` prod:

```
gaurav@gaurav-thinkpad:~/projects/ionic/vplanet-commerce$ ionic package build android --profile prod --release
The resources folder already exists. We will not overwrite your files unless you pass the --force argument.
Running with the force flag will overwrite your resources directory and modify your config.xml file
Preparing your resources...
splash ios Default-Portrait@~ipadpro.png (2048x2732) skipped, source image splash.png (2208x2208) too small
splash ios Default-Landscape@~ipadpro.png (2732x2048) skipped, source image splash.png (2208x2208) too small
Uploading your project to Ionic...
Submitting your app to Ionic Package...
Your app has been successfully submitted to Ionic Package!
Build ID: 5
We are now packaging your app.
```

You can now see from the list of package builds available, their build status, and download them by ID:

```
gaurav@gaurav-thinkpad:~/projects/ionic/vplanet-commerce$ ionic package list

 id | status   | platform | mode    | started                | finished
  5 | SUCCESS  | android  | release | Mar 9th, 2017 3:52:54  | Mar 9th, 2017 3:53:46
  4 | SUCCESS  | android  | debug   | Mar 9th, 2017 3:35:00  | Mar 9th, 2017 3:35:48
  3 | FAILED   | android  | debug   | Mar 9th, 2017 3:21:46  |
  2 | FAILED   | android  | debug   | Mar 9th, 2017 3:14:12  |
  1 | FAILED   | android  | debug   | Mar 9th, 2017 2:59:09  |

Showing 5 of your latest builds.
```

For downloading you need to just mention `BUILD_ID`:

```
$ ionic package download 4
```

Many times we did not apply the – –save flag while adding a cordova plugin, which did not make an entry inside `config.xml`. To make sure that the Ionic package service knows what plugins you need do a `ionic plugin save` so that all the plugin entries will be made in `config.xml`.

 Note: After the release of Ionic CLI 3 beta, commands like package, upload, setup, share and many other are removed. There commands will work fine with Ionic CLI v2 but if you are planning to upgrade so look for alternative commands in v3. Also, as v3 still in beta we don't have complete clarity about alternative to these commands.

Ionic View, Creator, and Playground

We have now covered almost all important Ionic Services, yet there are some Ionic tools available for Ionic applications that are really useful. Some of those tools are Ionic View, Creator, and Playground. Let's see how all these can help us.

Ionic View

Ionic View makes it easy to share your application with your team members or your clients without using TestFlight in iOS or beta testing on Android. We can share our application with anyone easily with complicated beta provisioning. The initial step is to upload our application, which will ask for the login credentials of your Ionic account:

Now you will be able to see your application with **ID f8fec798** in your portal. You can now share your application with the Ionic share command, as soon as you share your application with an account they will now be able to download and preview your application. Currently, there are around 27 plugins supported, which you can check out at `: https://docs.ionic.io/tools/view/`.

There could be a chance that your application doesn't fully work inside Ionic View as many Native plugins are not supported.

Collaborators or accounts with which I have shared my application with will be able to see the application in their app list and can simply click and preview the application. There are multiple options available such as clear data, sync latest version, or delete the app:

If you are not an owner, then you can enter the unique app ID and load the application. Once we are inside the application we have the option to send feedback to the developer by simply swiping up three fingers, which will open a feedback form page. As soon as you submit your feedback it will be visible in the **Feedback** tab in the dashboard:

Currently, feedback is only available to owners and collaborators.

Ionic creator and playground

There are another two tools that the Ionic creator and playground which Ionic platform provides. Ionic creator can be a good tool for prototyping and preparing UI mockups so that we can visualize quickly how the application will look like. Currently, under the free plan we can only drag and drop components, create pages/views, preview, share, and download code. Although, the Ionic team has launched many pro features that are available in paid plans such as code editing, theming, and packaging the creator application. Here we have tried to make a mockup for a currency convertor application using Ionic creator:

We have added three tabs from the components sections. With a simple drag and drop UI we can easily add items. On the right-hand side you will see specific page or component properties, which are selected from the left-hand side. For example, if we click on **List Page** on the left-hand side, then it will show properties for a page such as **Title**, **Routing URL**, **Background color**, and Left/Right header buttons.

Similarly, for each component we have different properties. There are some components and properties that are only available in the pro version:

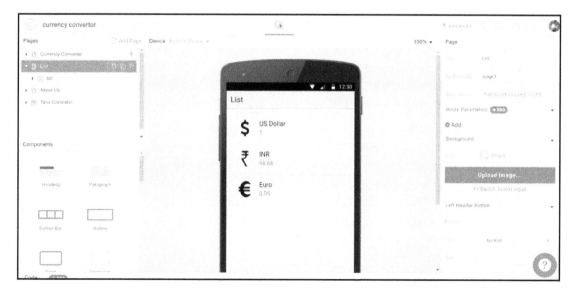

Let's see how the application will look like once we preview it. There are four device options available for preview, iPhone, iPad, Android phone, and Android tablet:

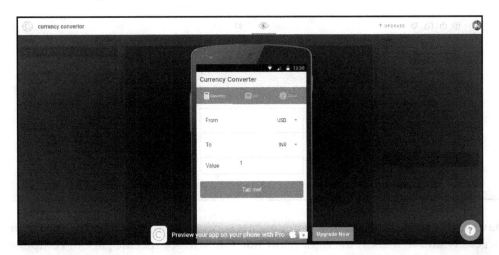

Ionic playground is a simple HTML, CSS, and JavaScript editor, which makes it incredibly easy to share Ionic demos with anyone around the world. You can think of playground as `jsfiddle` or `plnkr` for Ionic. We can fork and preview the applications we are working on. Playground can have a big impact on education and training developers in Ionic. For debugging purposes we can also use playground, another use case can be for quick prototypes and demos. Currently both Ionic creator and playground are based on Ionic 1 and our main focus here is on Ionic 3, so we will not look at these topics in too much detail. You can read more about Ionic Creator at `http://docs.usecreator.com/` and start playing with your first playground app at `http://play.ionic.io/`.

Ionic ecosystem comprises of Ionic market, Ionic jobs, Ionic forum, Ionic shop, and Ionic showcase. Let's get a brief idea about how exactly all of this works:

- **Ionic Market**: This is the place where you will find many Ionic themes and plugins available both free and paid. We can also submit our work on Ionic market if we have built some interesting theme or plugin that can be useful for other developers.
- **Ionic Jobs**: The best place to find a list of Ionic jobs.
- **Ionic Forum**: Here you can interact with other developers, ask questions, and share your demos and ideas. It's the best place to clear your doubts and check for similar problems you are facing.
- **Ionic Showcase**: You can submit your production deployed work here and share app stores links and screenshots.

Summary

With the amount of services and tools that the Ionic team is working on you can see that Ionic as a community and product are both growing at a good pace. We have discussed all Ionic ecosystem services starting from Ionic Cloud services and then the tools available for quick prototyping. Combining all these services can be really helpful; auth and Ionic DB together gives you a lot more options for your application to do customizations.

Ionic services such as IonicDB are a really interesting addition. There have been plans that new tools for testing will be launched by Ionic anytime in the future. The next big thing will be deployment of the same codebase as a desktop application using Electron bringing platform continuity.

In Chapter 5, *Authentication, Authorization, and Security* we will be looking into some security aspects for our Ionic application. Starting from some common security practices to integrating authentication and authorization inside our application and how we can protect our private views for accessing by non-authenticated users. We will look at storing sensitive data and read and write data through the network securely so that nobody can intercept it. Also, we will see some basic security key points that we should consider before publishing your application to app stores.

5
Authentication, Authorization, and Security

We will now be looking at another important aspect of building enterprise applications, which is security. There will be many applications that have user sensitive content and we don't want to expose everything to everyone. This is where authentication and authorization comes into play. Although, in this chapter, we will be more close to the Angular side of Ionic as the majority of work will be coding in Angular. Although, the Ionic community provides a good amount of information about it and how to use it, so we will be looking into all these things. Also, another important part in this chapter will be securing the application, as there are many things we have to take into consideration before deploying the app to production so that no one can easily get into your code.

During my research on securing hybrid applications and how it's different from native, I found this It doesn't matter if you use native code or hybrid code, if someone is dedicated enough, they can still work around it and see your source code. At the end it's our responsibility and how difficult we can make it to hack it around. Taking into consideration that we don't share or put sensitive information inside our codebase. We will be looking into all this in this chapter:

- Authentication
- Securing Ionic applications
- Demonstrating authorization in Ionic

Authentication

Angular is the core of Ionic and most of the work with respect to authentication and authorization will be handled in code. There are mainly two ways of implementing server side authentication for frontend applications. You look at the Ionic app as an Angular frontend application. All of what we will cover can be treated as common and can be used on the Web as well as Ionic-based hybrid applications.

Mainly there are two ways that we can usually handle authentication:

- Token-based
- Cookie-based

Token based authentication

Token based authentication is a newer approach of authentication in many web applications. In this approach authentication is successful when a user proves the server of their validity by passing the user token, which then is verified by the server. Before getting started with the token-based approach, let's quickly see some benefits of using this approach:

- Stateless
- Mobile-ready
- Performance
- More secure
- Standard based JWT
- There are many other benefits of using the token based approach, such as you can serve all your assets (HTML, CSS, and JS) via CDN and your server side is just an API, which will improve the performance

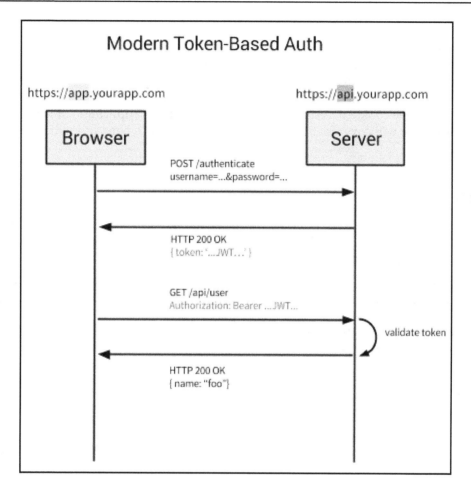

In a token based approach you first need to get a valid token, which you usually will get as a response of login request in many applications. Let's take an example of the Facebook connect API. Every time you login via Facebook it returns a token in the response. The token then will be used in future to get a set of user information from Facebook; Facebook uses that token to validate the user and on validation provides the information.

For an Ionic application, we can store that token to local storage or session storage for future use. We need to extend the default HTTP class for setting an authorization header for every outgoing request from our Ionic application. You can see how easy it's working with token based authentication; you don't need to think about security issues such as CSRF as you are not relying on cookies.

Almost all of the web applications previously were based on cookies and they handle user sessions in cookies. These require the server to authenticate the user cookies on every request. As we are moving towards a modern token based authentication approach, there are still some applications using a cookie-based authentication approach. Although we can implement the same cookie-based authentication on our Ionic application, it is not recommended. We will cover mainly the token based approach here and will demonstrate it in an application.

CORS

Cross Origin Resource Sharing (**CORS**) was designed mainly to access resources from cross domains. Cross-domain policy does not apply to Cordova, so we don't have to think much about when testing on a device:

```
XMLHttpRequest cannot load http://52.25.65.            /auth. No 'Access-Control-Allow-Origin' header is present on the requested resource. Origin 'http://localhost:8100' is therefore not allowed access.
```

While running the same app on a browser it might gave a CORS error while accessing resources from the cross domain. When you run the `ionic serve` command you will see this issue. There are four methods to bypass it:

- Disable Chrome security
- Browser extension
- Proxy server
- The easiest and most convenient way is to run the application in Google Chrome with `flag: --disable-web-security`:

```
$ google-chrome –disable-web-security
$ chromium-browser –disable-web-securityYou can run
    same command for chromium browser also. Make sure
    you have closed the running google-chrome or
    chromium before you ran it with this command. When
    you open it with disabling web security, you will
    see a warning message on the browser  mentioning
    "You are using an unsupported command-line flag: --
    disable-web-security. Stability and security will
    suffer".
```

Another way would be to install a browser extension, which will allow/disable CORS. Usually, these plugins internally change the response header for any XHR request when they are turned on. There are many extensions available, such as **CORS toggle** and **ForceCORS**.

Proxy server for the Ionic app

Setting up a proxy server will take some time and it is only required if you want to run `ionic serve` and `ionic run -l`. We will see how Ionic CLI provides an easily configurable proxy server.

Proxy server here will act as an intermediary for all the requests from client to server. It will take a client's request and issue a new request to the server and then on response from the server it will forward the response back to the client. Since the server is sending a new request to your destination, there will be no origin and therefore, no CORS needed. It is important to note that the browser adds in the origin header.

For setting up proxy server we have to do the following mentioned changes:

- Set up proxy in `ionic.config.json`
- Set up Angular constant
- Use constant in Angular services

Let's start with setting up proxy settings in the `ionic.config.json` file. Proxy settings contain two important things, `path` and `proxyUrl` inside the proxies array. You can have multiple proxies inside that array:

```
{
  "name": "ionic2-auth",
  "app_id": "",
  "v2": true,
  "proxies": [
  {
   "path": "/api",
   "proxyUrl": "http://yourwebsite.com /api"
  }
  ]
}
```

You now have specified inside your `ionic.config.json` that whenever you will access `http://localhost:8100/api` it will proxy that request to `http://api.yourwebsite.com/api` on your behalf. So this will solve the CORS issue. For setting up the APIs endpoints, we have created `constants` file where we have declared a constant, `API_URL`:

```
// src/providers/constants.ts
// While deploying application we switch to real URL
export const API_URL: string =
    'http://localhost:8100/api';
```

When building the application, we can replace it with a real `API_URL` that we mentioned as `proxyUrl`:

```
// src/providers/data-service.ts

import { Injectable } from '@angular/core';
import { HttpClient } from './http-client';
import { Observable } from 'rxjs/Observable';
import 'rxjs/add/operator/map';
import { API_URL } from './constants';

export class DataService {
 constructor(public http: HttpClient, public storage:
 Storage) {}

 login(data): Observable<any> {
 return this.http.post(API_URL + '/login', da-
 ta).map(res => res.json());
 }
}
```

You can directly import the constant file inside your `provider` class and call the login API. This way now you will complete the process of setting up the proxy for your Ionic application for development.

CSRF

Cross Site Request Forgery (CSRF) and many servers have this security feature. We can also keep this feature in our Ionic-based application. CSRF is a type of attack that occurs when malicious code performs an unwanted action on a trusted site mainly using the user's cookies itself when the user is authenticated.

To prevent CSRF with Angular and Ionic you need to set correct headers for CSRF using the specific cookie name. This will keep away hackers from sniffing your cookie session data and making requests pretending on behalf of an authenticated user.

This will set the CSRF request header to the current value of the CSRF cookie for any request type not in `allowedMethods`.

So to enable cookies based auth in Angular you have to take care of another important thing, setting the `withCredentials` flag of AJAX as true. This flag is set on a low level AJAX object, but in Angular we have to configure it with our `Http` method:

```
// Sample code snippet

import { Http, Headers, RequestOptions } from '@angular/http';
```

```
sendData() {
 let headers = new Headers();
 headers.append('X-CSRF-Token', token.token);
 let options = new RequestOptions({ headers: headers,
 withCredentials: true });
 this.http.post("URL", <stringified_data> , options)
}
```

While many developer face issues such as cookies expiring on restarting the application, if you face a similar issue, make sure your cookies have an expiry set. Actually, browsers clear cookies without an expiry on close.

Securing the Ionic application

With authentication and authorization you have already secured all your data, but there are still some things that we need to make sure of security while building your enterprise grade Ionic application. Many things here we will be working around Cordova to secure our application. As security is a deep and complicated topic, the most important thing is to stay updated and try using the latest versions of Cordova and Ionic. Usually, if there are some security vulnerabilities found in Cordova you will soon see a patch for that and it will get fixed.

Whitelisting

Domain whitelisting is a security model that controls the access to external domains. Cordova provides a configurable security policy to define which sites to access. For Cordova 4.0 and newer, we should use `cordova-plugin-whitelist` as it provides better security and configuration settings. Currently `cordova-whitelist-plugin` supports Android higher than 4 and for our platforms we have to configure security in the `config.xml` file: `<access origin="*" />`

While creating a project, by default all the domains are whitelisted, which means you will have access to all domains. Also, domain whitelisting is not available on and below Android API 10. So this is another important thing to note that you set your `min-target-sdk` above 10. Also, Ionic itself supports above Android version 4.4 . It will work on old devices and sdk, but it may have performance and other issues. Also, we don't have a big market share of devices with less than Android 4.4. So for building applications for the future we can ignore some share of market to make your application secure and good in performance. With new technologies coming, such as the BLE support application, even many native applications are only supporting latest versions.

We have already covered the whitelist plugin in `Chapter 3`, *Ionic Native and Plugins* so you already have an idea of how we can easily configure various setting and whitelist resources. The new content security policy in the plugin gives you control over which network requests (images, XHRs, and so on) are allowed to be made (via web view directly):

```
<!-- Allow everything but only from the same origin
  and foo.com -->
<meta http-equiv="Content-Security-Policy" con-
 tent="default-src 'self' foo.com">

<!-- Allows XHRs only over HTTPS on the same domain. -->
<meta http-equiv="Content-Security-Policy" con-
 tent="default-src 'self' https:">

<!-- Allow iframe to https://cordova.apache.org/ -->
<meta http-equiv="Content-Security-Policy" con-
 tent="default-src 'self'; frame-src 'self'
 https://cordova.apache.org">
```

You can find various other examples of configurations such as the previous ones in the official documentation of the whitelist plugin: `https://cordova.apache.org/docs/en/latest/reference/cordova-plugin-whitelist/`.

Enabling SSL

This is one of the most important points if you are transferring secure data over networks to servers, before uploading your application to various stores you should make it production ready and enable SSL for all the secure requests. As most of the Ionic applications use REST APIs to get or post data to servers, we should make sure that all of our secure requests are HTTPS. At the same time, if your application doesn't have any critical data then you can use HTTP also for this case.

Using self-signed certificates for production applications is not recommended. If you want to introduce an SSL, make sure that your server has a certificate properly signed by a well-known certificate authority. The reason is that accepting self-signed certificates bypasses the certificate chain validation, which allows any server certificate to be considered valid by the device. This opens up the communication to man-in-the-middle attacks. The data will be encrypted, but man-in-middle can decrypt it by the key by middleman. You can look at it as a normal web application where you always make sure that all the secure information is sent in encrypted form to the server so that no one can hack it in the middle. To make sure of this it is very important that you buy an SSL certificate from a well-known certificate authority.

Sensitive data outside the app

In our application, there might be some external APIs that we are using for fetching third-party content into our application. For this, you will get some kind of `secret_keys` specific to your account, which will be used when you hit their servers. Here we have to make sure that we don't place any of these keys inside your Ionic application as someone can easily extract that from your code base.

Here, the best way is to make a proxy request on your server or a custom request on your server-side code that will add the keys to the APIs calls. So you will be calling your server and then your server will pass a request to a third-party service with `secret_key` attached to the new request. In response the same response can be passed back to the client via a server from the third-party service:

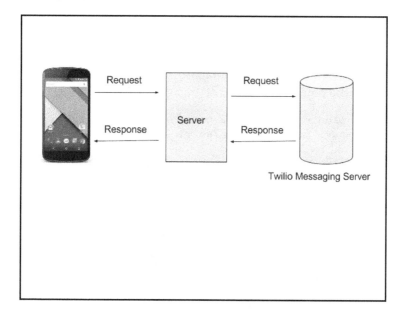

If, for example you have user specific keys then you have to ask the user to first login inside you application, which will then validate the user and the next API call from your server can get the respective `user_keys`, which can then be used for further communication with their third-party server. Also, make sure that you don't have any business login inside you application, which you can easily handle on the server side. Don't assume that your source is secure as Ionic applications that are Cordova-based are built from HTML and JavaScript assets packaged in a native container. Anyone can reverse engineer to get into your code, so that is why we don't ever place any sensitive information inside code.

Similarly, for BAAS services such as Ionic Cloud and Firebase, you have to place the keys on the client side for directly calling these platform APIs. Although, if we have proper access control and have protected the application with auth and login protected, then only authenticated users can make request to the APIs. This will secure it from hackers to exhaust your number of API calls for your account.

Secure storage

In Ionic application, many developers heavily depends upon localStorage or SQLite via Cordova plugin. Many times we have to store secret such as usernames, passwords, tokens, certificates, or any other sensitive information, In this case we can use secure storage Native plugin which uses keychain to encrypt data in iOS and in Android this plugin encrypts data to store it in SharedPreference. Inside an Ionic application, localStorage sandboxes the data only for your app. No, other application can access the data from outside, In case you are using SQLite as you application database, you can take advantage of SQLcipher plugin for encrypting the data inside SQLite:

```
$ ionic cordova:plugin add cordova-plugin-secure-storage
$ npm install --save @ionic-native/secure-storage
```

Let's see one such example can be storing user's mobile number:

```
import { SecureStorage, SecureStorageObject } from '@ionic-
native/secure-storage';

constructor(private secureStorage: SecureStorage) { }

 this.secureStorage.create('user_data')
 .then((storage: SecureStorageObject) => {

 storage.get('mobile_number')
 .then(
 data => console.log(data),
 error => console.log(error)
 );

 storage.set('mobile_number', '+917813123123')
 .then(
 data => console.log(data),
 error => console.log(error)
 );

 storage.remove('mobile_number')
```

```
.then(
data => console.log(data),
error => console.log(error)
);

});
```

Again, for all you are doing for encrypting the data we have to take are that we should never save passwords and other critical information inside application storage. Even in native applications it is not recommended because if someone is dedicated enough, they can still decrypt your data.

General security points

Most mobile operating systems provide hardware level encryption. This gives the devices extra protection when the device is locked. So we have to make sure that we always use the latest sdks and Ionic 2 only supports Android 4.4 and higher versions. Other than this there are some secure storage plugins for iOS and Android, which use keychain in case of iOS and for Android AES encryption. Another plugin is `PrivacyScreenPlugin`, which can be used for hiding sensitive information to be shown in app switchers in case of Android and iOS. Users will see splash screen in task switcher and if it is not detected it will show a black screen.

Reverse engineering is another concern of many developers who think that their application code can easily be extracted and repackaged in the application with some malicious code and uploaded back onto the Play Store. Although this same practice can be done in the case of Native applications, we can bypass this security concern in case of Ionic applications by downloading JavaScript at runtime and deleting it when the app closes. Although we need to think if we really require this as this will come with performance issues also. On the other hand, it's very difficult to do this in Java or objective-C and store your code base dynamically.

Obfuscation, minification, and compression are some of the ways you can secure your code within our Ionic application. We can use many open-source tools to minify and uglify your source before you package them to various app stores. Some popular tools are YUI compressor and UglifyJS, which can be used to uglify your JavaScript resources. Also, you should always minify all your JavaScript and CSS resources, which will improve the performance of your Ionic application and lower your APK file size. Some enterprise tools are also available such as Jscrambler, which does all kinds of minification and obfuscation using its proprietary solutions.

The following are some quick points that you can look into before you release your application for security context:

- Any sensitive business logic should be executed on the server
- Obfuscation and minification your JavaScript assets with uglify or use Jscrambler
- Do not use Android OS less than 4.0
- Use InAppBrowser for external links
- Validate user inputs for HTML forms
- Don't cache sensitive data
- Don't use `eval()` unless you know what you are doing
- Don't assume your source code is always secure
- Try to use keychain plugin for iOS

It's important to note that security is a critical topic and one should always be updated with new improvements in Cordova so we can improve our application in our next app release to app stores. Also, make sure that you don't lose any performance.

Demonstrating authorization in Ionic

Let's now build an Ionic application that will handle both authentication and authorization. Authorization refers to rules that determine who is allowed to do what. Authentication and authorization are central parts of securing an application:

Authentication = login + password (who you are)

Authorization = permissions (what you are allowed to do)

In this section, we will see how we can secure views of the Ionic application that are login protected. We will be working on the following points:

- `HomePage` will be public
- `ArticlesPage` is protected and will only be accessed after login
- Menu is also dynamically updated on auth status
- `LoginPage` will be inaccessible once a user is `loggedIn`
- Authorization header will be passed after login for each request

Let's start by creating the application and I have named it as `ionic2-auth`:

```
$ ionic start ionic2-auth sidemenu -v2
```

```
$ ionic platform add android
$ npm install @ionic/storage -save
$ ionic generate page articles
$ ionic generate page home
$ ionic generate page login
```

We will be using all the latest libraries and current Ionic version. Also, we now have spilt pane and responsive grid, which will make the Ionic applications closer to supporting desktop applications. We will mainly be using three views, HomePage, which is publicly accessible, ArticlesPage, which is protected, and LoginPage. HomePage we have set as rootPage as that is the public view. After this, let's look at the dynamic menu and login/logout events, which we will be handling inside app.component.ts:

```
// src/app/app.component.ts

import { Component, ViewChild } from '@angular/core';
import { Nav, Platform, MenuController, Events } from 'ionic-angular';
import { StatusBar, Splashscreen } from 'ionic-native';
import { ArticlesPage } from '../pages/articles/articles';
import { LoginPage } from '../pages/login/login';
import { HomePage } from '../pages/home/home';
import { DataService } from '../providers/data-service';
import { Storage } from '@ionic/storage';

@Component({
  templateUrl: 'app.html'
})
export class MyApp {
  @ViewChild(Nav) nav: Nav;

  rootPage: any = HomePage;

  pages: Array<{title: string, component: any}>;

  constructor(public platform: Platform,
    public storage: Storage,
    public data: DataService,
    public menu: MenuController,
    public events: Events
  ) {

    this.initializeApp();
    // used for an example of ngFor and navigation
    this.pages = [
      { title: 'Home', component: HomePage },
      { title: 'Articles', component: ArticlesPage }
    ];
```

```
      this.listenToLoginEvents();

      // decide which menu items should be hidden
      // by current login status stored in indexedDB
      this.data.isAuthorised().then((status) => {
        this.enableMenu(status === true);
      });

  }
```

Here we have initialized the application, set the menu pages, and checked the auth status so we can enable menu items accordingly:

```
  initializeApp() {
    this.platform.ready().then(() => {
      // Okay, so the platform is ready and our plugins
          are available.
      // Here you can do any higher level native things
          you might need.
      StatusBar.styleDefault();
      Splashscreen.hide();
    });
  }

  login() {
    this.nav.push(LoginPage);
  }

  logout() {
    this.events.publish('user:logout');
  }

  openPage(page) {
    // Reset the content nav to have just this page
    // we wouldn't want the back button to show in this
      scenario
    // Make sure to catch the error here else
    // you will get error
    this.nav.setRoot(page.component).catch(()=>{
      console.log("Page didnt load")
    })
  }

  listenToLoginEvents() {
    this.events.subscribe('user:login', (result) => {
      this.enableMenu(true);
      this.storage.set('user', result);
    });
```

```
    this.events.subscribe('user:logout', () => {
      this.enableMenu(false);
      this.storage.remove('user');
      this.nav.setRoot(HomePage).catch(()=>{
        console.log("Page didnt load")
      })
    });
  }

  enableMenu(loggedIn: boolean) {
    this.menu.enable(loggedIn, 'loggedInMenu');
    this.menu.enable(!loggedIn, 'loggedOutMenu');
  }
}
```

We have used the Events class here for our login and logout events, which is a publish-subscribe event system. As soon as the user has logged in on the LoginPage it will call the login event. We have stored the user object in Ionic storage, so later on for further HTTP requests we can use that to send authorization headers in requests. Also, we enable the specific menu here using the enableMenu() function:

```html
// src/app/app.html

<ion-split-pane>
  <ion-menu id="loggedOutMenu" [content]="content">
    <ion-header>
      <ion-toolbar>
        <ion-title>Menu</ion-title>
      </ion-toolbar>
    </ion-header>

    <ion-content>
      <ion-list>
        <button menuClose ion-item *ngFor="let p of pages"
          (click)="openPage(p)">
          {{p.title}}
        </button>
        <button menuClose ion-item (click)="login()">
          Login
        </button>
      </ion-list>
    </ion-content>

  </ion-menu>

  <ion-menu id="loggedInMenu" [content]="content">
    <ion-header>
```

```
          <ion-toolbar>
            <ion-title>Menu</ion-title>
          </ion-toolbar>
        </ion-header>

        <ion-content>
          <ion-list>
            <button menuClose ion-item *ngFor="let p of pages"
             (click)="openPage(p)">
               {{p.title}}
            </button>
            <button menuClose ion-item (click)="logout()">
               Logout
            </button>
          </ion-list>
        </ion-content>

      </ion-menu>

      <!-- Disable swipe-to-go-back because it's poor UX to
       combine STGB with side menus -->
      <ion-nav [root]="rootPage" main #content
       swipeBackEnabled="false"></ion-nav>
    </ion-split-pane>
```

In the preceding code is our `app.html` page where we have our `rootPage` and we have also used the split pane feature. You can see we have multiple ion-menus on the same side, which we can enable and disable according to the `ID` passed. Let's look into our `LoginPage` and `ArticlesPage` where we have called the APIs request and secured the routes:

```
// src/pages/login/login.ts

onLogin(form: NgForm) {
    this.submitted = true;

    if (form.valid) {
      // start Loader
      let loading = this.loadingCtrl.create({
        content: "Login wait...",
        duration: 20
      });
      loading.present();
      this.data.login(this.login).subscribe((result) => {
        this.events.publish('user:login', result);
        this.navCtrl.setRoot(HomePage);
        loading.dismiss();
        this.showToast(undefined);
      })
```

```
    }
  }

  ionViewWillEnter() {
    //Check if already authenticated
    this.data.isAuthorised().then(res => {
      if(res)
        this.navCtrl.setRoot(HomePage);
      else
        console.log("Please Login");
    })
  }
```

Here you can see we have called our `service` method for login and after that we have published the user: `Login` event so that we can update our application according to the auth status. The lifecycle event method `ionViewWillEnter()` is also used, where we have checked if the user is authorized and if it is then we don't route to the `HomePage`. Nav Guards are used to control entering and leaving from the views, which we will be using on `ArticlesPage`:

```
// src/pages/articles/srticles.ts

export class ArticlesPage {
  articles: Array<any>;
  constructor(
   public navCtrl: NavController,
   public navParams: NavParams,
    public toastCtrl: ToastController,
   public data: DataService) {
  }

  ionViewDidLoad() {
    console.log('ionViewDidLoad ArticlesPage');
    /*
      Fetching Data after View Loaded
      Don't put this request in constructor
      As, it will call the API irrespective of
      what ionViewCanEnter will return
    */
    this.fetchArticles();
  }

  fetchArticles() {
    this.data.getArticles().subscribe(articles => {
      this.articles = articles;
    })
  }
```

```
ionViewCanEnter() {
  console.log("in ionViewCanEnter")
  var self = this;
  return new Promise((resolve, reject) => {
    self.data.isAuthorised().then(res => {
      if(!res) {
        let toast = self.toastCtrl.create({
          message: "Please Login to access this view",
          duration: 1500
        });
        toast.present();
        reject(true);
      } else {
        resolve(true);
      }
    })
  });
}

}
```

Here inside the `ionViewCanEnter()` method we check if a user is authorized and accordingly it rejects or resolves promise. You will see that when a user is not `loggedIn` we will not be able to access the `ArticlePage`. In this case, even our APIs will not be called as we have called the `fetchArticles()` method inside the `ionViewDidLoad()` method, which is called only if `ionViewCanEnter` will be resolved:

```
// src/providers/data-service.ts

import { Injectable } from '@angular/core';
import { HttpClient } from './http-client';
import { Observable } from 'rxjs/Observable';
import 'rxjs/add/operator/map';

import { Storage } from '@ionic/storage';
import { API_URL } from './constants';

@Injectable()
export class DataService {

  constructor(public http: HttpClient, public storage:
  Storage) {}

  login(data): Observable<any> {
    return this.http.post(API_URL + '/users/login',
    data).map(res => res.json());
  }
```

```
getArticles(): Observable<any> {
  return this.http.get(API_URL + '/articles?
filter[limit]=10').map(res => res.json());
}

isAuthorised(): Promise<any> {
  return this.storage.get("user").then((val) => {
    console.log(val)
    if(val) {
      return true;
    }
  })
}
}
```

Inside data-service.ts we place our various APIs and the isAuthorised() method, which checks Ionic storage for user data and accordingly returns it back. Also, we will be using a http-client provider here for wrapping up our HTTP APIs calls. We use this to set the authorization header to every request after successful login:

```
// src/providers/http-client.ts

import { Injectable } from '@angular/core';
import { Http, Headers } from '@angular/http';
import { Storage } from '@ionic/storage';
import { Observable } from 'rxjs/Rx';

@Injectable()
export class HttpClient {

  constructor(public http: Http, public storage: Storage) {}

  buildHeaders(){
    let headers = new Headers({
      'Accept': 'application/json',
      'Content-Type': 'application/json; charset=utf-8'
    })

    return this.storage.get("user").then(
      (u) => {
        if(u){
          headers.append('Authorization', u.id);
          return headers;
        }
      }
    )
  }
```

```
get(url) {
  return Observable
      .fromPromise(this.buildHeaders())
      .switchMap((headers) => this.http.get(url, {
       headers: headers }));
}

post(url, data) {
  return Observable
      .fromPromise(this.buildHeaders())
      .switchMap((headers) => this.http.post(url, data,
       { headers: headers }));
  }
}
```

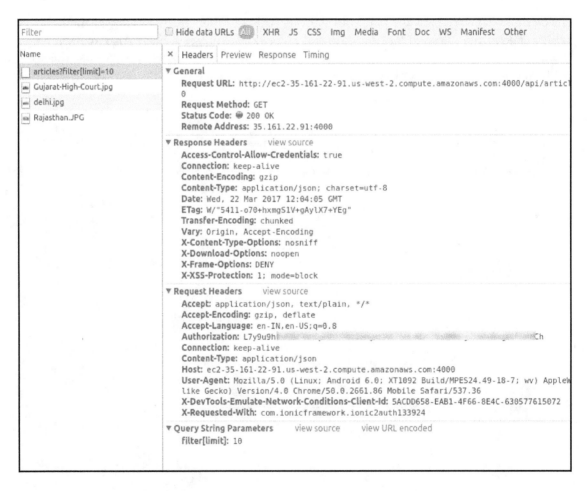

Here we have the `get()` and `post()` methods, which return observable, which further uses the `fromPromise()` method, which wraps another existing promise as an observable sequence. After this we have used the `switchMap()` method, which makes sure that the response comes in correct sequence as we want the headers and on that basis we set it to the `Http` method.

 Observable here is being brought in from `rxjs/Rx` instead of from `rxjs/Observable`. This will automatically get us access to the `fromPromise` method, otherwise it will give an error.

For building headers, our `buildHeaders()` method checks the Ionic storage and if the token is present it appends it to authorization headers. We have covered all important aspects of securing our application. We can anytime extend it further if there are roles returning from the backend. Another good addition here can be extending `HTTP` class to add some interceptor for `401` requests so it logs the users out from the application if the user token is expired.

Summary

When it comes to securing your application, the most important thing is to properly secure your server-side APIs. So this way we will not lose any critical data and other's data will not get shared publicly.

Although, when we think about securing our code base, then nothing is 100% in native as well as hybrid mobile development. The goal should be to get as close to that number as possible though. Start by limiting your exposure and taking care that no sensitive data and logic should be on the client side. You should make sure you are doing your best to secure your hybrid mobile apps. However, if someone is dedicated enough, they can still work around it.

We now have covered some important concepts of security inside and outside of your Ionic application. We will now move one step ahead with integrations of our Ionic app with BasS services such as Firebase and Ionic Cloud. In case you have very simple requirements for you Ionic application and don't have much server logics you can quickly build an Ionic application with these services without setting up any server-side code base. We will be building a **TasteBite** application where we will be rating food items and restaurants combined, so that users can know what the famous dishes in their cities are and where they can find it cooked best.

6
TasteBite App with Firebase

Till now we have covered many things in theory with respect to Ionic applications. We are half-way into this exciting journey developing high-performance enterprise applications. We now will focus on building an entire application from start to end and will be using a backend as a service platform for Firebase. We will be developing a food rating application where the user can search for the best food items by multiple users in his city. Also, we will allow authenticated users to give reviews and ratings to the food items, which will help us rate food items inside our application. Firebase is one such BAAS platform and we will be demonstrating it in this chapter; other than Firebase we can also use Stamply, Kinvey, or Backendless, which are similar platforms that help frontend developers build application rapidly. After this brief introduction about our application, let's overview some important topics we will be covering:

- Introduction to Firebase and TasteBite apps
- Authentication and security using Firebase
- CRUD and securing data with AngularFire2

Introduction to Firebase and TasteBite apps

Getting started with Firebase is really simple; that is the main reason many startups and growing products use all these Backend as a Service platforms like Firebase. We will quickly start by creating a TasteBite application with Firebase and Ionic. In this application we integrated an AngularFire2 wrapper for Angular applications using Firebase, although, we can simply use the Javascript SDK also inside our application. You need to create an account at Firebase (initially you can have a free account; as your consumption increases you can upgrade your account easily).

Firebase has made the life of developers easier as they don't have to spend time building backend server to build simple APIs. We can easily do all CRUD operations with minimum effort and with Web socket supports it's super fast in syncing data. Also, Firebase offers offline sync, user authentication, and features for securing the data with sets of rules:

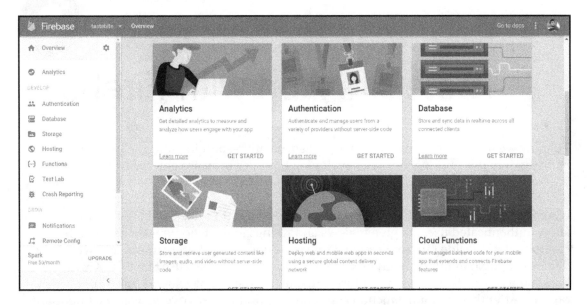

As, soon as you create an account you create an app inside Firebase; you will find an option to add a project. After you add/create a project you can now manage and enter your applications. You will see many options such as **Analytics**, **Authentication**, **Database**, **Storage**, **Hosting**, **Notification** and many other options. We will use many of them in our application as we move forward creating data for our application. You can already see in the following screenshot that we have some data stored in our application. All Firebase database data is store as JSON objects and no relational tables exist.

Building TasteBite and installing Firebase

Let's first start by creating an application from scratch and install all the plugins and packages required in this application. In this application we will be using the Tabs template and integrating Lazy Loading. We will be using geolocation, camera, and in-app browser plugins in our application:

```
gaurav@gaurav-thinkpad:~/projects$ ionic start tastebite tabs
✓ Creating directory /home/gaurav/projects/tastebite - done!
✓ Downloading 'tabs' starter template - done!
✓ Updating project dependencies to add required plugins - done!
✓ Creating configuration file for the new project - done!
✓ Executing: npm install within the newly created project directory - done!

♬ ♩ ♬ ♩  Your Ionic app is ready to go! ♬ ♩ ♬ ♩

Run your app in the browser (great for initial development):
  ionic serve

Run on a device or simulator:
  ionic cordova:run ios

Test and share your app on a device with the Ionic View app:
  http://view.ionic.io

? Link this app to your Ionic Dashboard to use tools like Ionic View? No

Go to your newly created project: cd /home/gaurav/projects/tastebite

gaurav@gaurav-thinkpad:~/projects$
```

```
    $ ionic start tastebite tabs
$ ionic cordova:platform add android
$ ionic cordova:plugin add cordova-plugin-inappbrowser
$ ionic cordova:plugin add cordova-plugin-geolocation
$ ionic cordova:plugin add cordova-plugin-camera
$ npm install --save @ionic-native/in-app-browser
$ npm install --save @ionic-native/geolocation
$ npm install --save @ionic-native/camera
```

Now, we are ready to setup Firebase in our Ionic application:

```
$ npm install angularfire2 firebase –save
```

Next we will import the angularfire2 library in app.module.ts:

```
// src/app/app.module.ts

import { AngularFireModule } from 'angularfire2';

export const firebaseConfig = {
  apiKey: "AIzaSyDnd6hMcfJDWGdDr6EOSS05NoMM2CeY2bw",
  authDomain: "tastebite-ec5c4.firebaseapp.com",
  databaseURL: "https://tastebite-
  ec5c4.firebaseio.com",
  storageBucket: "tastebite-ec5c4.appspot.com",
```

```
      messagingSenderId: "762350093850"
};


@NgModule({
  declarations: [
    Tastebite
  ],
  imports: [
    BrowserModule,
    HttpModule,
    IonicModule.forRoot(Tastebite, {
      preloadModules: true
    }),
    AngularFireModule.initializeApp(firebaseConfig)
  ],
```

Make sure you include `firebaseConfig` as shown in the previous example, which will have multiple configuration parameters. Also, the next step will be to inject `firebaseConfig` while initializing our Firebase module. We are now ready to read and write data inside our application.

Authentication and security using Firebase

We will now look into Firebase user authentication and how we can secure our data. Firebase supports login and authentication internally and also provides support for integrating with social login providers. Most applications need some kind of login mechanism internally; with Firebase it's super easy to set this up. Whenever we login using Firebase a session is created and a unique ID `uid` is returned by Firebase, which is distinct across all providers and never change for a specific authenticated user. This will also help us later in securing our data.

Once the user is authenticated its session is managed and remembered even after the application is restarted. Firebase provides user authentication for multiple platforms:

- Custom - for integrating with existing authentication systems
- E-mail - register and authenticate users by e-mail and password.
- Anonymous - for guest accounts without any personal information required
- Facebook
- Google
- Twitter
- GitHub

Enabling providers

We need to enable auth provider from our application authentication **Sign-In Method** tab. We have enabled e-mail based login and Facebook login for our TasteBite app application. It's really easy to sign up, log in and log out users with e-mail-based login. We can customize auth if we already have some authentication systems; if we want to have guest login we can use anonymous user authentication. We will demonstrate how easily we can integrate social logins also.

All social logins require a client ID and client secret for the application created on social channels. All social providers have different ways of creating an application and after creating an application we can put the client ID and client secret inside our Firebase app. We need to make sure that we have placed the Oauth redirect URL inside the Facebook login product: `https://tastebite-ec5c4.firebaseapp.com/__/auth/handler`. Before this you have to create an application on Facebook whose `app_id` and `app_secret` we will add back to our Firebase Facebook provider settings. After this your backend work is done and you just now have to call a function from the Ionic application.

We will now generate `AuthServiceProvider` which we will use for AngularFire2 authentication:

```
$ ionic g provider AuthService
```

Before we start working on this provider we need to set up `FirebaseAuthConfig` inside our `app.module.ts` file:

```
import { AngularFireModule, AuthProviders, AuthMethods } from
    'angularfire2';

const myFirebaseAuthConfig = {
  provider: AuthProviders.Facebook,
  method: AuthMethods.Redirect
};

imports: [
  AngularFireModule.initializeApp(firebaseConfig,
    myFirebaseAuthConfig)
]
```

This way we don't need to pass this `config` to the `AngularFireAuth` based login function. Although we can override this configuration and use any other provider by passing a login method internally. In the code above you can see `AuthMethod` is selected as `Redirect`, but for testing on browsers you can change it to `Popup`. Let's look at our `AuthService` now:

```
import { Injectable } from '@angular/core';
```

```
import { AngularFireAuth, FirebaseAuthState } from 'angularfire2';
import * as firebase from 'firebase';

import { Platform } from 'ionic-angular';
import { Facebook, FacebookLoginResponse } from '@ionic-
   native/facebook';


@Injectable()
export class AuthServiceProvider {

  private authState: FirebaseAuthState;

  constructor(public auth$: AngularFireAuth, private fb:
  Facebook,
     public platform: Platform) {
    this.authState = auth$.getAuth();
    auth$.subscribe((state: FirebaseAuthState) => {
      console.log(state);
      this.authState = state;
    });
  }

  get authenticated(): boolean {
    return this.authState !== null;
  }

  signInWithFacebook(): firebase.Promise<FirebaseAuthState |
     FacebookLoginResponse> {
    if (this.platform.is('cordova')) {
      return this.fb.login(['email',
  'public_profile']).then(res
        => {
        const facebookCredential =
           firebase.auth.FacebookAuthProvider.credential
           (res.authResponse.accessToken);
        return firebase.auth()
           .signInWithCredential(facebookCredential);
      });
    } else {
      return this.auth$.login();
    }
  }

  signOut(): void {
    this.auth$.logout();
  }
```

```
authData(): any {
  if (this.authState != null) {
    return this.authState.auth;
  } else {
    return '';
  }
}

}
```

We initially installed in app browser Cordova plug into handle this authentication and redirects.

We have also checked that, if we are on a real device, we use the Ionic native plugin for Facebook. Here we first call the Facebook plugin, and after a successful login, and a token is received from Facebook, we pass it to the Firebase `firebase.auth.FacebookAuthProvider` function for authentication. We have used Facebook login in our TasteBite application, although if we want to we can implement other providers also.

User auth state

There are some auth-related APIs available with Firebase that help us get the auth state and persist it. To list any changes in the auth with regard to any change in password or session expiration, we have to listen the subscribe function of the `AngularFireAuth` class when you initialize your application:

```
constructor(public auth$: AngularFireAuth) {
    this.authState = auth$.getAuth();
    auth$.subscribe((state: FirebaseAuthState) => {
      console.log(state);
      this.authState = state;
    });
}
```

We have placed this inside our `AuthServiceProvider`; similarly if you want to stop listening to the changes you can call the `offAuth()` function and you will not get any updates. Firebase saves a session inside local storage till the user session is valid, although if you need to get auth you can call the `getAuth()` function and get all the session details. For low requests on the server we can use the `localstorage` method to get all details from storage:

```
signOut(): void {
    this.auth$.logout();
```

```
}
```

Users are currently always persisted until `logout()` is called (or the user clears local storage). Finally, we need to have a function to log out the session from Firebase; we can use `unauth()`.

CRUD and securing data with AngularFire2

We now have covered auth-related code and set up the building block for our TasteBite application. Next, we will look into Firebase **Realtime Database,** which is a NoSQL cloud database. Data is stored as JSON and synchronized in real time using web-sockets. The Firebase database persists to the disk, which gives our application offline support also. We will be using here two objects, `foodItems` and `rateItems`, where `foodItems` will have details about the food dish and `rateItems` is a nested object that contains ratings and reviews for each dish + city combination:

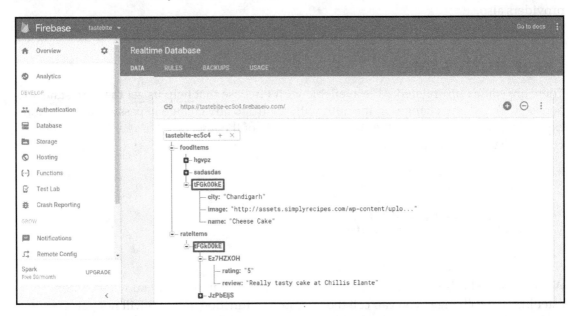

As you can see how we have organized the `foodItem` data with respect to userid so we can differentiate user-related data. We can traverse to the `personalFavourite` and then to user data clicks on the dashboard. We can access any of the child nodes in the data. We can visualize the data in the dashboard; also Firebase team has created a Chrome plugin called Vulcan, which we can use to see the data tree.

As, data stored inside Firebase is all objects and does not provide any native support for arrays. However, to ease development Firebase provides methods in its SDK that push data to an array. Firebase stores the data with sequential keys:

```
// push send this
['a', 'b', 'c', 'd', 'e']

// Firebase databases store this
{0: 'a', 1: 'b', 2: 'c', 3: 'd', 4: 'e'}
```

There are restrictions with the use of array type storage on Firebase. Also, with Firebase you can have depth of child node till 32 level and the key length can be 768 bytes. We will not go into deep about Firebase basis as we can easily get that on its official documentation.

CRUD operations

To read or write data from the database, we need an instance of database reference. Although we are using `AngularFire` so these functions are wrapped and we use the function provided by `AngularFire` library. Sometimes, we have to use low-level methods also like we will be doing later in the project. Let's now first start querying the `foodItems` to show that on our `HomePage`:

```
// src/pages/home/home.ts

import { Component } from '@angular/core';
import { IonicPage, NavController, NavParams } from 'ionic-angular';
import { AngularFire, FirebaseListObservable }  from 'angularfire2';
import { CityProvider } from '../../providers/city/city'

@IonicPage()
@Component({
  selector: 'page-home',
  templateUrl: 'home.html',
})
export class HomePage {

  foodItems: FirebaseListObservable<any[]>;
  item;
  constructor(
    public navCtrl: NavController,
    public navParams: NavParams,
    public cityService: CityProvider,
    public af:AngularFire) {
```

As we will be querying data by city, we are using `cityService`, which returns our current location's city according to our geolocation. After this we will only get the data that is from our `city`:

```
this.cityService.getCityName().then(result => {
  let cityName = result[0]['address_components'][3].long_name;
  this.foodItems = af.database.list('/foodItems', {
    query: {
      orderByChild: 'city',
      equalTo: cityName
    }
  });
})
}

ratingPage(data:any) {
  console.log(data);
  // In FoodRatingPage we will be showing
  // all the reviews and rating for each item
  this.navCtrl.push("FoodRatingPage", data);
}

}
```

Next, if the user clicks on any of the food item it will show all the reviews and rating associated with that food item.

As mentioned earlier, we are using a city provider which first uses the Ionic Native geolocation plugin and then passes the latitude and longitude to the Google API, which returns use entire detail of that place from street name, the city, state and country. We then parse `city` name from this and use it in our application:

```
// src/providers/city/city.ts

import { Injectable } from '@angular/core';
import { Http } from '@angular/http';
import 'rxjs/add/operator/map';

import { Geolocation } from '@ionic-native/geolocation';


@Injectable()
export class CityProvider {

  constructor(public http: Http, private geolocation:
  Geolocation) {
    console.log('Hello CityProvider Provider');
```

```
    }

    getCityName() {
      return new Promise(resolve => {
        this.geolocation.getCurrentPosition().then((resp) => {
          let url =
            "http://maps.googleapis.com/maps/api/geocode/json?latlng="
            +resp.coords.latitude+","+resp.coords.longitude
          this.http.get(url).map(res =>
          res.json()).subscribe(data => {
            resolve(data.results);
          });
        }).catch((error) => {
          console.log('Error getting location', error);
        });
      })
    }

}
```

When the user traverses the to next page, we pass the specific `foodItem` data, which is later used here to fetch `foodRatings`. We store `foodRatings` and `foodItems` with the same `$key` so it's easier to fetch data in this case:

```
// src/pages/food-rating/food-rating

import { Component } from '@angular/core';
import { IonicPage, NavController, NavParams } from 'ionic-angular';

import { AngularFire, FirebaseListObservable } from 'angularfire2';


@IonicPage()
@Component({
  selector: 'page-food-rating',
  templateUrl: 'food-rating.html',
})
export class FoodRatingPage {
  item:any;
  ratings:any;
  foodRatings: FirebaseListObservable<any[]>;

  constructor(
    public navCtrl: NavController,
    public navParams: NavParams,
    public af: AngularFire) {

    this.item = this.navParams.data;
```

```
        this.foodRatings =
            af.database.list('/rateItems/'+this.navParams.data.$key);
    }

    ionViewDidLoad() {
      console.log('ionViewDidLoad FoodRatingPage');
    }

    ratingPage() {
      // Will Push to RateFoodPage where we will rate this
      specific
          foodItem+city combo
      this.navCtrl.push("RateFoodPage",
      this.navParams.data);
    }


  }
```

Now, let's see how we can create an entry in Firebase. Say we have to give a **review** and **rating** to `foodItem`, we do that on our `rateFoodPage`. Here again we use the same `$key` to store a new review in the database.

```
rateItems
  ⊟── tFGk00kE
      ⊟── Ez7HZXOH
          ├── rating: "5"
          └── review: "Really tasty cake at Chillis Elante"
      ⊟── JzPbEljS
          ├── rating: "3"
          └── review: "Not that good taste in Gopals Chandigarh"
```

```
// src/pages/rate-food/rate-food

import { Component } from '@angular/core';
import { IonicPage, NavController, NavParams } from 'ionic-
   angular';

import { AngularFire, FirebaseListObservable } from 'angularfire2';
import * as firebase from 'firebase';

@IonicPage()
@Component({
```

```
    selector: 'page-rate-food',
    templateUrl: 'rate-food.html',
  })
export class RateFoodPage {

    rateItems: FirebaseListObservable<any[]>;
    data: {name?: string, review?: string, rating?: any} = {};
    constructor(
      public navCtrl: NavController,
      public navParams: NavParams,
      public af: AngularFire) {

      this.rateItems = af.database.list('/rateItems');
      console.log(this.navParams.data);
      this.data.name = this.navParams.data.name;
    }

  rate() {
    let data = {
      "rating": this.data.rating,
      "review": this.data.review
    }
    let url = "rateItems/" + this.navParams.data.$key +
    "/" +
        this.makeid();
    firebase.database().ref().child(url).set(data);
    this.navCtrl.pop();
  }

  makeid() {
    var text = "";
    var possible =
        "ABCDEFGHIJKLMNOPQRSTUVWXYZabcde
        fghijklmnopqrstuvwxyz0123456789";

    for( var i=0; i < 8; i++ )
        text += possible.charAt(Math.floor(Math.random() *
            possible.length));

    return text;
  }
}
```

We have used the `makeid()` function to generate random `$keys` for reviews/rating of the same `foodItems` by multiple users. Here you can see we haven't used the `AngularFire` method and rather used directly Firebase method because we will not be able to insert data in this way using `AngularFire` standard functions.

Other than creating and querying data, we can update and delete objects using `AngularFire`. We haven't used these in our application but it is really easy to perform these functions:

```
// Deleted the object and handles error via promise
const promise = af.database.object('/item').remove();
promise
  .then(_ => console.log('success'))
  .catch(err => console.log(err, 'You dont have access!'));

// updates data of an object
const itemObservable = af.database.object('/item');
itemObservable.update({ age: newAge });
```

This is one small example of how quickly we can prepare prototypes using Firebase. We can further extend this application with uploading images and integrating the push and analytics services of Firebase. There are multiple possibilities available and from here you can easily start your journey.

 There have been regular updates to every technology and framework so, I have to make sure that the code-base linked to each chapter will be updated as often as possible. Also, many times all features can't be written down in the chapter so I have tried my level best to cover as many topics in the demonstration projects with in each chapter. So, please make sure that you go through the code-base and wherever there is a inconsistency I will add a comment in the code.

Structuring and securing data

While storing and structuring data inside Firebase we have to think twice about how we will be fetching the data for the application. This problem is the same as when we use any other NOSQL database such as MongoDB. We should prefer flattening data rather than nesting everything and while fetching the data we have to join the flattened data.

Let's take the example of a blogging website where users create blog posts and all blog posts have comments. So, rather than nesting data in Firebase we prefer to divide it:

```
"users": {
    "user1": {
      "name": "John Smith",
      "type": "editor"
      "blogs": {
        "post1": {
          "id": 123456
```

```
            "title": "welcome of ionic",
            "post":  "Ionic is opensource SDK or hybrid
             apps"
          },
          "post2": {
            "id" : 123457
            "title": "welcome of ionic",
            "post":  "Ionic is opensource SDK or hybrid
             apps"
          }
        }
      },
      "user2": {
        "name": "Adam",
        "type": "admin"
        "blogs": {
          "post1": {
            "id": 234567
            "title": "welcome of ionic",
            "post":  "Ionic is opensource SDK or hybrid
             apps"
          }
        }
      }
    }
  }
```

This way of storing the data is initially easy to handle but later on, as the data increases, querying the dataset will be slow and performance will be harmed. So, the best way to store data is:

```
"users": {
    "user1": {
      "name": "John Smith",
      "type": "editor"
      "blogs": {
        "123456" : true,
        "123457" : true
      }
    },
    "user2": {
      "name": "Adam",
      "type": "editor"
      "blogs": {
        "234567" : true
      }
    }
  },
  "blogs": {
```

```
      "123456": {
        "title": "welcome of ionic",
        "post":  "Ionic is opensource SDK or hybrid apps"
      },
      "123457": {
        "title": "welcome of ionic 2",
        "post":  "Ionic is opensource SDK or hybrid apps"
      }
    }
  }
```

Now we have stored the data inside Firebase and the next step is to fetch the data and joining back. Joining flattened data is a bit difficult but this approach comes with a lot of performance and modularity. Let's see how we can fetch the above data and join it.

As Firebase uses web sockets performance will be high while fetching the data this way and Firebase internally do optimizaition on incoming and outdoing requests.

Firebase also provides a set of security rules for securing our data and help determine who has read and write permission for the data set. All the security rules in dashboard inside the **Security & Rules** tab. Inside our application our application data is user-defined and associated to the user, so we have set the rules accordingly:

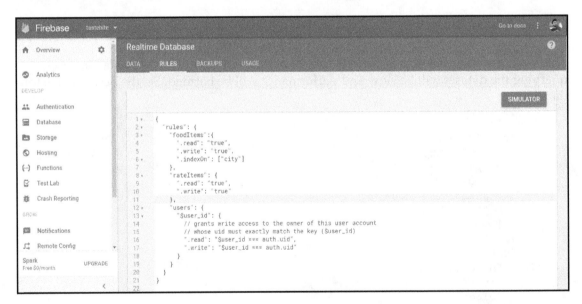

There are three types of rule:

- `.read` - when data is allowed to read be by users
- `.write` - when data is allowed to be written
- `.validate` - defines correctly formatted data and validates before pushing to the database

Security rules are in the JSON structure as you see in the previous screenshot and this makes it easier to understand and set rules. You will d associated to user, so we have set the rules accordingly. In our application we first need to store the user-related data during the signup process such as the name, e-mail, and other details:

```
{
    "rules": {
      "users": {
        "$user_id": {
          ".read": "$user_id === auth.uid",
          ".write": "$user_id === auth.uid"
        }
      }
    }
}
```

We have created a `users` model to store these details. We have nested the data inside a `$user_id` key so that we can create an access control and only user who created this data set can edit and access it. Similarly for a `"feeds"` users can only read the feeds and can't write a feed on it:

```
"feeds": {
    ".read": true,
    ".write": false
}
```

There are ways inside Firebase that help increase performance while querying data. For example if we have score of students inside a model and we need to query that data by score then we can use `.indexOn` inside the rule so that performance while fetching records increases. You can study more about securing data inside Firebase here: `https://firebase.google.com/docs/database/security/`.

Summary

Using Ionic with the BASS platform really ease a lot of tensions for a startup company as you will have the application built in less time for multiple platforms and without needing any resources or time for set up the backend server. Nowadays, a lot of growing companies are moving towards Ionic due to its development speed and lower cost. Small teams of developers with good JavaScript skills can build everything and even if they want to have a backend for custom tasks they can use Node.js and build an entire stack over JavaScript. Platform such as Firebase is really powerful for mobile application as it uses web sockets, which are really fast and provide real-time data synchronization.

Next we will work on some interesting latest technology IOT. Ionic has been quite famous with developers for quickly preparing applications to control various IOT devices via bluetooth or HTTP APIs. We will focus on building applications using the BLE and iBeacons protocols. We will be also going through Google's Physical Web initiative using the Eddystone protocol, which gives physical interaction with physical objects. Also, there are multiple IOT devices available on the market and we can interface with them easily with Ionic applications and solve many problems.

7
Ionic, IOT, and Beacons

This chapter will be one of the most interesting chapter as we will be using Ionic and later BLE technology to build an application. IOT devices have been in the news for quite some time and the amount of data we are getting from these devices opens up an entire new space. We can already see how the fitness band market has evolved in recent times and now people have already building various other IOT devices or gadgets to make human life easier. On the other hand, we need applications to capture and process the data from devices. Developers are already started working on the data sets and adding machine learning to suggest health and fitness tips to the users, according to the health stats they capture. Thus, this industry is growing at a fast pace and Ionic is easing the learning curve to enable many developers to create mobile applications.

Until now we have covered a lot many things with respect to Ionic applications and we are almost near end of this exciting journey for developing high performance enterprise applications. We will now be focusing more on what we can build next in IOT space with Ionic applications. We will be discussing the following topics:

- Ionic and IOT
- BLE and the physical Web
- Proximity-based screen lock using BLE

Ionic and IOT

Developers have started to understand the potential of IOT, which is enormous. With so many different IOT devices in the market, there is also a need for applications to control these devices remotely. Now here comes Ionic to the rescue to help developers create apps quickly and inexpensively. Ionic is really efficient for building these apps and at the same time offers full native support to devices features for example, access to Bluetooth so sending signals to IOT devices. Already there are multiple devices that are connected over Bluetooth or HTTP APIs for data syncing and performing various actions. There are so many exciting projects currently going on and we will try to learn more about them. Let's look into some of the reasons for using Ionic 2 for building your next IOT application:

- With Ionic 2, we can have Web apps, Android, iOS and soon desktop with a single code base. Specially, for a prototype purpose its really helpful
- Ionic apps are built-in web technologies so they are really quick to get started
- Ionic comes with multiple set of tools so you can efficiently bootstrap your application and deploy it easily

Ionic + Node = deadly combination

Node.js is solving many problems for hardware and IOT devices. There have been so many ways we can use Node.js to build the backend part which can talk with the Ionic app over HTTP APIs. One such example is where Ionic, Node.js, MQTT, and Wi-Fi module is used and we have a mobile app controlled relay through the Wi-Fi module over Internet. We are using the free online broker, MQTT for sending message from the Node.js server to the Wi-Fi module which are set up in Arduino IDE. On the other hand, we are using HTTP Rest APIs to call the Node.js server from the Ionic app. So, this way we will be able to send signals to Wi-Fi router and can perform specific actions.

Users need some interface to interact with these IOT devices, which open the path for Ionic-based mobile applications. We will be implementing one such node module, bleno, for implementing BLE peripherals later in the chapter. Noble and bleno are two such modules used for BLE implementation. Nobile scans the peripherals around its proximity and bleno transmits as a peripheral. Also, other than this there are multiple IOT modules available as NPM packages, which will get you started really quickly:

- **CyclonJS**: mainly used for robotics and IOT
- **NodeRED:** tool for wiring together multiple IOT hardware devices
- **DeviceJS**: Javascript engine for IOT, Jquery for IOT
- **noduino**: Node.js framework for controlling Arduino controls

Websockets can be another messaging protocol just like MQT. We will be using websockets in this chapter in our demonstration. There are endless possibilities that we can work on. Arduino and Raspberry Pi are two famous boards used for building applications which can interact with other hardware devices. There are multiple examples you will find online to build during hackathons or for learning purposes where developers turn on an LED using Ionic application in real time. One such example is with Raspberry Pi GPIO pins and breadboard connected to it. Node.js code is residing on Raspberry Pi and an LED bulb connected on a breadboard. Another example we can think of is to connect with sensors such as temperature, GPS, and so on. You can find an interesting example of Ionic applications using an MQTT for fetching data from sensors in real time and visualizing in the application at `https://github.com/arjunsk/Ionic-IOT-Monitor`.

The possibilities are endless and as the IOT industry is growing you will find more sensors and hardware devices coming up for solving many real-life problems. You can even think of controlling all your home devices and building a home automation project, similar to what Mark Zuckerberg did some months back with his Jarvis AI . Although he used AI for his project we can simply start with just controlling all our smart devices according to our needs. The best part as a developer here is that you would love it a lot even when you write only a small piece of code that can make your life easier.

BLE, beaconsm, and the physical Web

BLE stands for Bluetooth Low Energy, Bluetooth 4.0 has have brought innovative changes to the IOT industry. There was time when Bluetooth was considered as a medium to exchange files but now, as technology has evolved we can already see how the fitness band and smart-watch market has grown. Bluetooth smart devices consume low energy which helps the smartphone battery last longer and proves to be effective for the IOT industry. With Bluetooth once again penetrating into the market. we are seeing high usage of Bluetooth devices inside the IOT industry.

iBeacons

Beacons are devices broadcasting small data packets at regular interval. This is a one-way communication method and is based on BLE technology. Beacons advertise themselves so that devices such as smartphones can interact with them and can perform actions such as sending a push notification. Beacons enable you to get information or content according to your location and are widely effective indoors where GPS doesn't work perfectly. We can think of a beacon as a lighthouse whose sole motive is to send out signals and is unaware of devices nearby. On the other hand, smartphone devices receive these signals similar to ships at sea. Beacons just transmit some sort of ID but the data behind it is stored over the cloud and we can get the content from the cloud by sending a reference ID. Similarly you can think like the captain of a ship and can now refer to the maps to ascertain where exactly they are passing the reference from the lighthouse.

Let's look at the benefits end users can have from different applications of beacons:

- Context-specific information
- Indoor navigation
- Contactless payments
- Real-time offers
- Personalized experience

Beacons technology enables customers to receive the right information at the right time. Macy's mall recently installed beacons on multiple locations and is already reaping the rewards and building stronger relationship with customers:

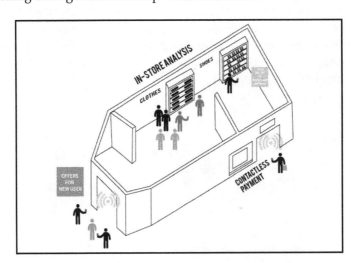

You can see from the preceding illustration how we can interact with beacons inside a shopping mall. We can place multiple beacons on multiple locations inside the shopping mall. As soon as a user enters the shopping mall, he will receive a notification with the latest offers available inside the mall.

There has been a lot of growth in the e-commerce sector and people now prefer shopping online for their ease and comfort. Physical stores are facing the heat due to this and that is why they have to bring innovation to their retail stores to attract customers and get sales. One such way is to bring IOT devices to the physical retails stores and start interacting with shoppers' smartphones to bring the latest updates to them:

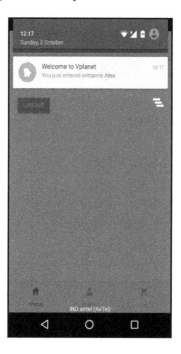

This is a screenshot of an application which I built during a hackathon, which uses Kontakt beacons for broadcasting iBeacons and our Ionic-based mobile application detects it and triggers a local notification.

We have been talking a lot about beacons, let's look into what are iBeacons are an iBeacon is a protocol developed by Apple, which sets a layout about how to transmit data signals. Generally, there are two protocols for BLE broadcasting iBeacon and eddystone. Standard iBeacon advertisement consists of UUID, major and minor:

```
UUID: f7826da6-4fa2-4e98-8024-bc5b71e0893e
```

```
Major ID: 65279

Minor ID: 63150
```

Eddystone and the physical Web

Eddystone is a BLE beacon protocol released by Google which is open source and cross-platform. Eddystone contains three types of data frame types: Eddystone-UID, Eddystone-URL, and Eddystone-TLM:

- **Eddystone-UID**: A unique 16-byte static ID, typically used by native apps
- **Eddystone-URL**: A physical web project than transmits a compressed URL which is directly usable by the client
- **Eddystone-TLM**: Broadcasts beacons information such as battery level, sensor data and other information

Almost most of the major beacons manufacturer now supports both iBeacons and Eddystone data formats. Google is now promoting beacons technology so to provide better location and proximity experiences. Currently, eddystone-url frame is widely used in physical Web initiative.

The physical Web is an open approach to enable quick and seamless interactions with physical objects and locations.

The physical Web extends the power of the URL to everyday physical objects. This enables you to interact with any physical device such as a vending machine rental without the need to download the device specific application. It's so smooth that you will just tap the URL attached to the device and start using it inside your web browser. The physical web is a really small discovery service that gives specific URLs which nearby mobile devices can receive. This will lead you to a fully interactive web page or deep line to a native application.

We are already facing the challenge of installing multiple applications which take up our memory and storage of devices. Again, managing multiple applications will be a headache and this is another reason why developers are thinking about the mobile web with PWA. Similarly, as the IOT devices keep on increasing, installing an app for each device is not scaling. So, here the physical Web is helping out; you can just open the URL and interact with the devices on the Web, which in the backend can talk with the device over multiple protocols. Let's think of real-life example. If you are giving a talk, you can use beacon to broadcast the URL for the slides and ask the audience to turn on Bluetooth and the location to receive the URL.

We also don't need a real hardware beacon device to demonstrate this or broadcast during a conference. There is an NPM package available, Eddystone-beacon, which we can use to advertise an Eddystone URL:

```
$ npm install eddystone-beacon -save
$ sudo node -e "require('eddystone-
beacon').advertiseUrl('https://www.facebook.com/veloice');
```

Just make sure that your Bluetooth is enabled on the Linux laptop which you are using. Now enable Bluetooth and the location on your Android mobile and you will see a notification with the title and description of the web URL:

Here I have broadcast three different URLs and you can see the details of all three URLs in the notification menu. This is a small example of how we can share URLs in public. You can find many such examples on physical Web website (https://google.github.io/physical-web/) and see how you can build something out of this initiative.

We can perform various functions with iBeacons and Eddystone such as ranging, region monitoring, and other settings. Based on tx and RSSI values, we can get the approximate distance between devices and beacons. We can even do indoor mapping and geofensing with the help of beacons.

Proximity-based screen lock using BLE

This is something that I have been thinking of for a while to implement and is a real-life problem for me. It's time to look into how practical it can be to use BLE stack with an Ionic application. While working in my office, I have to quite often had leave my seat for taking some urgent calls or to attend some meetings. Most of the time I forget to screen lock my laptop. So, I thought of building this proximity based screen lock application using BLE and Websockets.

Let's first look quickly look into how the application will work. Ionic based application listens to the iBeacon packets from the continuously broadcasting laptop. As, soon as I move from my seat with my mobile, at some point it will exit from the broadcasting region of my laptop. Immediately it will send a socket request over the network to the Node.js based application running on my Ubuntu laptop and call the screen lock command on my laptop. Similarly, as I come back in to within the broadcasting range it will send another socket request to unlock the laptop.

The technology stacks that we will be using here are:

- Node.js
- Web Sockects
- BLE
- Ionic
- Bash script

We have to prepare our Node.js backend and bash script first, to get started with our application. Before this, make sure you have all the `debian` packages installed for working on BLE and Node.js.

```
$ sudo apt-get install bluetooth bluez libbluetooth-
  dev libudev-dev
```

Now we can `init` our project using the `npm init` command:

```
gaurav@gaurav-thinkpad:~/projects$
gaurav@gaurav-thinkpad:~/projects$ mkdir screen-lock
gaurav@gaurav-thinkpad:~/projects$ cd screen-lock/
gaurav@gaurav-thinkpad:~/projects/screen-lock$ npm init
This utility will walk you through creating a package.json file.
It only covers the most common items, and tries to guess sensible defaults.

See `npm help json` for definitive documentation on these fields
and exactly what they do.

Use `npm install <pkg> --save` afterwards to install a package and
save it as a dependency in the package.json file.

Press ^C at any time to quit.
name: (screen-lock)
version: (1.0.0)
description: Proximity based screen lock application
entry point: (index.js)
test command:
git repository:
keywords:
author: Gaurav Saini
license: (ISC) MIT
About to write to /home/gaurav/projects/screen-lock/package.json:

{
  "name": "screen-lock",
  "version": "1.0.0",
  "description": "Proximity based screen lock application",
  "main": "index.js",
  "scripts": {
    "test": "echo \"Error: no test specified\" && exit 1"
  },
  "author": "Gaurav Saini",
  "license": "MIT"
}


Is this ok? (yes) yes
gaurav@gaurav-thinkpad:~/projects/screen-lock$
```

We are using the following NPM packages for our application:

- `app` and `http` - for creating a web server and listening on port 5000
- `io` - for websocket implementation
- `bleno` - for implementing BLE peripherals
- `child_process` - for executing, `bash` file

```
$ npm install express http bleno socket.io
```

```
// index.js

var app = require('express')(),
    http = require('http').Server(app),
    io = require('socket.io')(http),
    bleno = require('bleno');
    exec = require('child_process').exec;

// Start advertising iBeacon
var uuid = '38a48d7e-05ab-11e7-93ae-92361f002671';
var major = 10; // 0x0000 - 0xffff
var minor = 5; // 0x0000 - 0xffff
var measuredPower = -59 // -128 - 127

bleno.on('stateChange', function(state) {
  console.log('on -> stateChange: ' + state);
  if (state === 'poweredOn') {
    bleno.startAdvertisingIBeacon(uuid, major, minor,
      measuredPower);
  } else {
    bleno.stopAdvertising();
  }
});
```

Initially, we are declaring the properties for the advertising the iBeacon. We are listening to the stateChange listener here so in case Bluetooth is ON/OFF we can start or stop the advertisement of iBeacon with the startAdvertisingIBeacon() and stopAdvertising() methods respectively:

```
// Create websocket
io.on('connection', (socket) => {
  console.log('user connected');

  socket.on('disconnect', function(){
    console.log('user disconnected');
  });

  socket.on('status', (data) => {
    console.log(data);
    if (data == "lock" || data == "unlock") {
      // Executing bash file for lock and unlock
      exec(__dirname + '/unlock.sh ' + data,
        function(error,
          stdout, stderr) {
          var out = error ? stderr : stdout;
          socket.emit('result', data)
          console.log(out);
        });
```

```
        } else {
          console.log("Please pass correct input ", data);
        }
      });

    });

    http.listen(5000, () => {
      console.log('started on port 5000');
    });
```

We will be talking between our Ionic application and Node.js code via websockets. So, as soon as data is pushed from the mobile application, it will check the status as `lock` or `unlock` and accordingly call the bash script. Following is the `unlock.sh` bash script we have used. Make sure you make the file permission as executable:

```bash
#!/bin/bash
export XAUTHORITY=/home/$USER/.Xauthority
export DISPLAY=:0.0

case "$1" in
    unlock)
        echo "Unlocking"
        loginctl unlock-sessions
        xset dpms force on
        xset s reset
        ;;
    lock)
        echo "Locking"
        loginctl lock-sessions
        xset dpms force off
        ;;
esac
```

Let's test our application and run the `index, js` file. We have to run our file as a `root` user as it grants node privileges for Bluetooth:

```
gaurav@gaurav-thinkpad:~/projects/experiments/lock-ubuntu$ sudo node .
started on port 5000
on -> stateChange: poweredOn
user connected
lock
Locking
```

Now we are ready to start with our Ionic application to interact with the backend. Let's start the application using tabs starter template:

```
gaurav@gaurav-thinkpad:~/projects/ionic$ ionic start lock-app tabs --v2
Creating an Ionic 2.x app in /home/gaurav/projects/ionic/lock-app based on the tabs template.

Downloading: https://github.com/driftyco/ionic2-app-base/archive/master.zip
Downloading: https://github.com/driftyco/ionic2-starter-tabs/archive/master.zip
Installing npm packages (may take a minute or two)...
/
♬ ♪ ♬ ♪ Your Ionic app is ready to go! ♬ ♪ ♬ ♪

Some helpful tips:

Run your app in the browser (great for initial development):
  ionic serve

Run on a device or simulator:
  ionic run ios[android,browser]

Share your app with testers, and test on device easily with the Ionic View companion app:
  http://view.ionic.io
gaurav@gaurav-thinkpad:~/projects/ionic$
```

As we are already aware of what all we will be using in our application, let's install the platform, Cordova plugins, and NPM packages:

```
// add android platform
$ ionic platform add android

// add ibeacon and diagnostic plugins
$ionic plugin add cordova-plugin-ibeacon --save
$ionic plugin add cordova.plugins.diagnostic -save

// install NPM package for socket.io-client
$ npm install socket.io-client --save
```

We will get started initially by checking if Bluetooth is enabled and asking the user for location permissions, which has become mandatory since Android 6 when using diagnostic plugins:

```
// src/app/app.component.ts

import { Component } from '@angular/core';
import { Platform } from 'ionic-angular';
import { IBeacon, Diagnostic, StatusBar, Splashscreen } from
    'ionic-native';

import { TabsPage } from '../pages/tabs/tabs';

declare var cordova: any;
```

```
@Component({
  templateUrl: 'app.html'
})
export class MyApp {
  rootPage = TabsPage;

  constructor(platform: Platform) {
    platform.ready().then(() => {
      // Okay, so the platform is ready and our
        plugins
        are
        available.
      // Here you can do any higher level native
        things
        you might
        need.
      StatusBar.styleDefault();
      Splashscreen.hide();

      if (platform.is('cordova')) {

        // Checks if bluetooth is enabled
        IBeacon.isBluetoothEnabled().then((isEnabled) => {
          console.log("isEnabled: " + isEnabled);
          if (!isEnabled) {
            // If not, enableBluetooth() enables it
            IBeacon.enableBluetooth();
          }
        })

        // Check if location is enabled
        Diagnostic.isLocationEnabled().then((result)
         => {
          // If not, open location settings
          if(!result) {
            Diagnostic.switchToLocationSettings();
          }
        })

        // Diagnostic plugin
        // After Android 6, we need to explicitly
        // ask for permission from the user
        Diagnostic.requestRuntimePermissions([
          cordova.plugins.diagnostic
          .permission.ACCESS_FINE_LOCATION,
          cordova.plugins.diagnostic.
           permission.ACCESS_COARSE_LOCATION
        ])
```

```
                    }
                });
            }
        }
```

Initially, we check `isBluetoothEnabled()`, which returns a Boolean according to the status. After that, if it returns `false`, then we call the `enableBluetooth()` method to enable Bluetooth. Similarly, `isLocationEnabled()` is called to check initially if the location is enabled and, if not, call the `switchToLocationSettings()` method, which opens the location settings.

`Diagnostic.requestRuntimePermissions()` is required to get runtime permissions, which is required for all devices running on Android 6 and over. For the monitoring and ranging of BLE, we need ACCESS_FINE_LOCATION and ACCESS_COARSE_LOCATION permissions:

Now we will set up iBeacon monitoring and the `Socket.IO` connection inside our home page:

```
// src/pages/home/home.ts
```

```
import { Component } from '@angular/core';
import { NavController, Platform } from 'ionic-angular';
import { IBeacon } from 'ionic-native';
import { Http } from '@angular/http';
import * as io from 'socket.io-client';
import 'rxjs/add/operator/map';

@Component({
  selector: 'page-home',
  templateUrl: 'home.html'
})
export class HomePage {

  socket: any;
  socketHost: string = 'http://192.168.0.103:5000';
  laptop_status: any;

  constructor(
    public http: Http,
    public navCtrl: NavController,
    public platform: Platform) {

    this.initialize();

    if (platform.is('cordova')) {

      // For iOS 8 we have to request permission
         explicitly
      IBeacon.requestAlwaysAuthorization();

      // create a new delegate and register it with the
         native
         layer
      let delegate = IBeacon.Delegate();

      delegate.didStartMonitoringForRegion()
        .subscribe(
          data =>
      console.log('didStartMonitoringForRegion: ',
            data),
          error => console.error()
        );

      delegate.didEnterRegion()
        .subscribe(
          data => {
            console.log("Entered Region", data);
            this.manageLaptop(data, "unlock");
```

```
              }
          );

        delegate.didExitRegion()
          .subscribe(
            data => {
              console.log("Exited Region", data);
              this.manageLaptop(data, "lock");
            }
          );

        let beaconRegionEntrance =
        IBeacon.BeaconRegion('laptop',
            '38a48d7e-05ab-11e7-93ae-92361f002671', 10, 5);


    IBeacon.startMonitoringForRegion(beaconRegionEntrance)
          .then(
            () => console.log('Started Monitoring',
                    beaconRegionEntrance),
            error => console.error('Failed to begin
              monitoring: ',
                error)
          );

        IBeacon.enableDebugNotifications();
      }
    }

  manageLaptop(data, status) {
      this.socket.emit('status', status);
  }
```

The `startMonitoringForRegion()` method starts the monitoring for the specified region with the `uuid`, major and minor, which our Node.js code is broadcasting. Now, as you move away from your seat and move out from the Bluetooth range, the `didExitRegion()` method will be called, which is subscribed to further call our `manageLaptop()` method with the data and lock status. Similarly, when we get back, `didEnterRegion()` will be called:

```
    initialize() {
      this.socket = io.connect(this.socketHost);

      this.socket.on("connect", (msg) => {
        console.log('on connect');
      });
```

```
this.socket.on("reconnecting", (msg) => {
  console.log('on reconnecting');
});

this.socket.on('disconnect', function() {
  console.log('user disconnected');
});

this.socket.on("result", (msg) => {
  this.laptop_status = msg;
  console.log(msg, "==received=msg====")
});
  }

}
```

Inside our `manageLaptop()`, method we emit our custom event to the server. Before that we have also `initialize()` our function to make the initial connection with the server on the host where our backend Node.js app is running.

This way our entire demonstration is completed. Personally, for me it worked out really well and, with the manual button provided for locking and unlocking, I can also lock my laptop remotely. You can work on many small hacks and prototypes that can be built quickly with Ionic and IOT.

Summary

Beacons and their applications will be emerging soon in multiple sectors , particularly in the retail sector. Yet, it's really new technology as far as wide-spread implementations are concerned. Retailers should try to think of new innovations with new technologies so the user-experience can be enhanced. We have discussed and implemented many use cases of how we can use IOT with Ionic applications quickly. I have been following Ionic blogs regularly and seen so many teams using Ionic for interfacing with their hardware devices, one such is jewelbots which enables when a friend is nearby or send you an message. Opportunities are enormous, its just need an idea in mind to implement it and take it to the world.

We are almost near completion of our journey, but we have another interesting topic still waiting to be covered next. We discussed in this chapter about the problem of installing multiple native applications for accessing devices, so this same problem is solved by the introduction of PWA support for Ionic. PWA has made the user experience immersive, even on low Internet speeds. Many e-commerce and large tech companies in India have launched their progressive web application after Flipkart launched it initially with collaboration with the Chrome team. All of a sudden, we can see the market is growing and that developers build platforms for making current web application to enable PWA features. Although, it will not end the native application market as we don't have the power of all the plugins. The best part of using Ionic is that the exact same code can be converted to both a PWA and a native application.

8
Ionic + PWA = Magic

The next big thing coming after mobile applications are progressive web applications for mobile web. **Progressive web applications (PWA)** was originally started by Google in 2015 and we can already see many major companies came up with their PWA version for mobile web. Before PWA, many developers and companies lost interest in mobile web and were just focusing on native applications, but Google became strict and started penalizing sites for having banners to download native applications or not having mobile-friendly site. So after this developers started working on mobile web also and at the same time Google Chrome launched a support for PWA, and Flipkart initially came up with their progressive web application working closely with the Chrome team.

It's all about delivering amazing user experiences, whether it's a native application or a PWA, end users don't care as long as they have a smooth experience. We will be looking at the following topics in this chapter:

- What and why PWAs?
- Ionic supports PWA
- Offline currency converter PWA
- Future of PWAs

What and why PWA?

PWAs are a new breed of web applications. They combine the benefits of a native application with the low friction nature of the Web. Progressive web applications start off as simple websites, but as the user interacts with them, they progressively gain super powers. They transform from a website into something much more, such as a traditional native application.

PWAs are:

- **Reliable**: PWAs load instantly and never show the downasaur, even in bad network conditions. When an application is launched from the user's home screen, service workers (will be discussed later) enable a progressive web application to load it instantly, regardless of the network state.
- **Portable**: There are many strategies for native application portability (including hybrid applications), but unlike those strategies PWAs don't change your deployment and packaging model.
- **Engaging**: PWAs feel like a natural application on the device, with a great user experience. They are installable, they can be added to the user home screen, without the need for any application store. They offer full screen experience like a native application and even re-engage users with web push notifications.
- **Fast**: PWAs are fast and respond quickly to interactions with smooth animations. Users always expect sites to be fast and smooth. According to a survey, 53% of users will abandon a site if it takes more than three seconds to load.

Advantages of PWA

PWAs have certain advantages over native applications, it's the future for the next billion users who will come online in the coming few years. Now we will talk about the advantages and why they are crucial to mobile users:

- **Install prompt**: Install banners are really crucial and they increase the engagement with the application. Currently, this can't be controlled by developers, but Chrome decides when to prompt the user to show the option of **Add to Home Screen**.
- **Works offline**: Internet connections can be really poor in many areas, which is why we need offline support for our application. PWAs help us gain better user experience and engagement. Cache APIs and `IndexedDB` are the two recommendations for enabling offline access to your PWA.
- **Lightweight**: Progressive web applications are just web pages and really lightweight and quick to install. PWAs do not consume large space for installation and are updated in real time, as they don't have to go via Play Store approvals.
- **Linkable and discoverable**: We can easily share URLs and is identifiable as an application due to manifest and service worker registration scope. Also, allows search engine to find it.

- **Web push notifications**: Service workers also help in handling the web push notifications, which help to re-engage users. PWA are served only over HTTPs, so they ensure security.
- **Update on use**: Unlike the usual native application update strategy, PWAs are updated like web pages: using them gets you the latest version.
- **Analyze before installation**: You don't need to install them to start using them. They are simple web pages. Users choose to **install** when they want to.

Disadvantages of PWA

The following are the disadvantages of PWA:

- **Traffic through store**: PWA miss significant traffic who use Play Store or application store for their primary search. Users can be redirected to Play Store to download the native application using mobile web, but it doesn't work in case of PWA.
- **Plugin integration**: Plugins such as Facebook and Google login cannot fetch data from their applications, we need to fetch it separately on the Web.
- **Promotion**: It is getting difficult to promote PWA experience on social media as companies are making their own in-application browsers.
- **Browser support**: Full support for PWAs are not available in some browsers varying on devices.

Ionic supports PWA

The Ionic team are great fans of the progressive web application, which offers an application-like experience using modern web APIs. Since Ionic 3 is built with web technologies that work perfectly in web browsers, this gives you a huge advantage when building your PWA with Ionic, because if you know how to build an Ionic 3 application, then you can easily build a PWA in a short period of time, you just have to add service workers and a web manifest file. We will be building an offline currency converter with PWA support and will finally host it on Firebase hosting.

What makes a web application a PWA?

In simple words, a PWA is meant to be eventually installed to the user's home screen, which means they will access it exactly like a native application. So users expect the same experience as when using a native application. Since Ionic 3 is designed to create a native application experience with standard web technologies, this is already handled.

PWA should be very performant, since PWA is going to be installed to the home screen right along side with native applications. Ionic 3 is built from the ground with performance in mind, which means you barely have to think about it. As PWA is going to be installed on the user's home screen, this process requires a web manifest for your application. Thankfully, Ionic 3 CLI comes with a built-in manifest file.

The majority of native applications works launch offline and give a limited offline experience to the user, so that a PWA can do for the user. With the help of service workers, we can now make our web application work while offline.

PWA with Ionic

The two things that are needed for a web application to become a progressive web application are a mainfest and a service worker. Ionic 3 now offers both of these by default with every new application that gets created.

The manifest gives the browser metadata about your application, so that when the user chooses to add your PWA to their home screen, it should know what icon to use, how it should be displayed, the name of the application, and more. By default, Ionic provides a manifest that's already linked to `index.html`. The manifest file can be found in the `www` folder of the Ionic project.

The service worker script allows you to control how your PWA will work and fetch resources. Service workers give your PWA functionality, such as the ability to work offline, sending push notifications, adding to user's a home screen, background sync, and more. The service workers are now provided by default for every new Ionic application that is created. We will be covering manifest and service workers script in the next section.

Web manifest and service workers are the two things that are needed to make a web application a progressive web application. Now we will be discussing these two about how we can include in our application.

Web manifest

When you build an Ionic application, Ionic CLI automatically generates a manifest file inside `src` folder with the name `manifest.json`. This file is responsible for how our application will look like on a mobile screen when installed. Ionic automatically sets the path inside your `index.html` file:

```
<link rel="manifest" href="manifest.json">
```

So you don't have to worry about the path, you just have to change the `manifest.json` file according to your needs and Ionic will take care of the rest. As you can see, we have several options available (`name`, `short_name`, and so on) in our manifest file, we will now be discussing the manifest file used in our currency converter application:

```
// src/manifest.json

{
  "name": "Currency Converter PWA",
  "short_name": "Currency",
  "start_url": "index.html",
  "display": "standalone",
  "icons": [{
    "src": "assets/logo.png",
    "sizes": "512x512",
    "type": "image/png"
  }],
  "background_color": "#db0000",
  "theme_color": "#000000",
  "gcm_sender_id": "124917710568"
}
```

- `fullscreen`: All of the available user area is used and no user agent Chrome is shown.
- `display`: Defines the developers preferred display mode for the web application.
- `start_url`: Specifies the URL that loads when a user launches the application from a device. In our Ionic application it will always be `index.html`.
- `short_name`: This is to be used where there is insufficient space to display the full name of the web application.
- `name`: This is what the user will see when they open the application on the splashscreen page.
- `Standalone`: The application will look and feel like a `standalone` application. This can include the application having a different window, its own icon in the application launcher, and so on.

- `minimal-ui`: The application will look and feel like a standalone application, but it will have a minimal set of UI elements for controlling navigation. The elements will vary by browser.
- `Browser`: The application opens in a conventional browser tab or new window, depending on the browser and the platform. This is the default.
- `Icons`: Array of icons used in various different places for the application, for example, when we add it to the home screen this icon is placed with the `short_name` of the application.
- `background_color`: Expected background color for the web application.
- `theme_color`: Default theme color for an application.
- `gcm_sender_id`: Used for sending web push notifications, represents `sender_id` from Firebase Cloud:

You can see how the application splash screen will look like when we have installed it, with some of the preceding properties implemented.

Service workers

Using service workers with Ionic 3 is really easy, Ionic comes with the `service-worker.js` file, which is initially commented in the `index.html` file. We just need to uncomment that section and you will see service workers registered in your browser.

Service workers essentially act as proxy servers that sit between web applications and between the browser and network (when available). They are intended to (amongst other things) enable the creation of effective offline experiences, intercepting network requests, and taking appropriate action based on whether the network is available and updated assets reside on the server. They will also allow access to push notifications and background sync APIs.

A service worker has life cycle and events for various actions. We initially start by registering a service worker on our page, on which successful registration life cycle initiate with install and activate events. Ionic 3 comes with the built-in registration code in the `index.html` file inside the `src` folder, all you need is to uncomment the code. Now you can check that a service worker is enabled by going to `chrome://serviceworker-internals/` by looking at your site:

As you can see, our locally served application at `localhost:8100` has a registered service worker. Although, when we deploy PWAs, it must be served securely (HTTPS without cert errors). Firebase hosting and GitHub pages host over HTTPs, so they can be best place to host your the PWA application.

Now that we are done with the registration, it's time to install events to start its work. Inside our install event we cache all our static files. The `event.waitUntil()` method ensures that the service worker is not installed until the code inside it is not executed. Now, `cache.addAll()` caches the static resources in cache storage:

```
var cacheName = 'cache-v1';
// pre-cache our key assets
var files = [
  './build/main.js',
  './build/main.css',
  './build/polyfills.js',
  'index.html',
  'manifest.json'
];

self.addEventListener('install', (event) => {
  console.info('Event: Install');

  event.waitUntil(
    caches.open(cacheName)
    .then((cache) => {
      //[] of files to cache & if any of the file not present
&grave;addAll&grave; will fail
      return cache.addAll(files)
        .then(() => {
          console.info('All files are cached');
          return self.skipWaiting(); //To forces the waiting service worker
to become the active service worker
        })
        .catch((error) => {
          console.error('Failed to cache', error);
        })
    })
  );
});
```

 After sw-toolbox support we don't need to use this as it automatically does this with its precache method, which we will use in our currency converter demonstration.

After the installation is completed, the service worker activates itself and `EventListener` is called. We use this method for cache management, let's say that in the future you a have new version of a service worker file and older cache is no longer required, so here we clean the old caches that we don't require now:

```
/*
```

```
    ACTIVATE EVENT: triggered once after registering, also used to clean up
caches.
*/

//Adding 'activate' event listener
self.addEventListener('activate', (event) => {
  console.info('Event: Activate');

  //Remove old and unwanted caches
  event.waitUntil(
    caches.keys().then((cacheNames) => {
      return Promise.all(
        cacheNames.map((cache) => {
          if (cache !== cacheName) { //cacheName = 'cache-v1'
            return caches.delete(cache); //Deleting the cache
          }
        })
      );
    })
  );
});
```

Now we have a service worker installed and running. We need to fetch resources and
according to different requests we return the responses. In the following snippet, we have
used the event.respondWith() method, which will wait until we check whether a
request is available in cache or not. If a request is not cached, we add it to the cache:

```
self.addEventListener('fetch', (event) => {
  console.info('Event: Fetch');

  var request = event.request;

  //Tell the browser to wait for newtwork request and respond with below
  event.respondWith(
    //If request is already in cache, return it
    caches.match(request).then((response) => {
      if (response) {
        return response;
      }

      //if request is not cached, add it to cache
      return fetch(request).then((response) => {
        var responseToCache = response.clone();
        caches.open(cacheName).then((cache) => {
          cache.put(request, responseToCache).catch((err) => {
            console.warn(request.url + ': ' + err.message);
          });
        });
```

```
            return response;
        });
      })
   );
});
```

 After sw-toolbox support we use sw-toolbox router methods for fetching the resources and caching them according to different strategies available, which we will look at further in our demonstration.

Service workers can perform push events, to create user notifications, to open a specific page, or to run some background script:

```
self.addEventListener('push', (event) => {
console.log('Received a push message', event);

const title = 'Yay a message.';
const body = 'We have received a push notification.';
const icon = 'assets/icon.png';
const tag = 'simple-push-demo-notification-tag';

event.waitUntil(
  self.registration.showNotification(title, {
    body: body,
    icon: icon,
    tag: tag
  })
);
});
```

Recently, Ionic added support for sw-toolbox, which is a `service-worker` library and it reduces a lot of work.

Service worker toolbox (**sw-toolbox**) is now built-in to the starters and conference application. This allows you to customize your service worker using a high-level API instead of using the raw service worker API. Our out-of-the-box configuration will give your application a good, network independent experience, but you can easily customize this to fit your application's unique use cases.

We have gone through the various events of service workers and how exactly they work. An important thing here is to use the Chrome developer tools for debugging service works and look into the servicer worker status and caches available. For further research on service workers and what all we can do using them you can look into the following documentation links by Google and Mozilla:

```
https://developer.mozilla.org/en-US/docs/Web/API/Service_Worker_API/Using_Servi
ce_Workers
https://developers.google.com/web/fundamentals/getting-started/primers/servi
ce-workers.
```

Offline currency converter PWA

We have looked into what exactly PWA and service-workers are, now let's build a PWA application using Ionic. We will be implementing an offline currency converter, which according to the network state fetches the latest data from `fixer.io` or from the cache or storage data. The application will have a listing of currency conversion rates according to base currency.

Firstly, we will start a new Ionic 3 application using Ionic CLI and install the NPM package for `@ionic/storage`:

```
gaurav@gaurav-thinkpad:~/projects/ionic$ ionic start currency-converter-pwa tabs --v2
Creating an Ionic 2.x app in /home/gaurav/projects/ionic/currency-converter-pwa based on the tabs template.

Downloading: https://github.com/driftyco/ionic2-app-base/archive/master.zip
Downloading: https://github.com/driftyco/ionic2-starter-tabs/archive/master.zip
Installing npm packages (may take a minute or two)...
\
♬ ♩ ♬ ♩  Your Ionic app is ready to go! ♬ ♩ ♬ ♩

Some helpful tips:

Run your app in the browser (great for initial development):
  ionic serve

Run on a device or simulator:
  ionic run ios[android,browser]

Share your app with testers, and test on device easily with the Ionic View companion app:
  http://view.ionic.io
gaurav@gaurav-thinkpad:~/projects/ionic$ cd currency-converter-pwa/
```

```
$ npm install --save @ionic/storage
```

We will be running it using Ionic serve, which will host it on the `8100` port. Now as discussed in our last section, we will uncomment the service worker code in the `index.html` file in the `src` folder.

In our application, we have mainly two views, `HomePage` and `ListPage`. Other than that we will be using two provider classes, `currency-service` and `storage-service`, for calling the exchange rate APIs and for storing data in indexDB, respectively:

```
$ ionic generate page list
$ ionic generate provider currency-service
$ ionic generate provider storage-service
```

Let's start with our `index.html` file where we have registered `service-worker` and also subscribed to `pushManager` for sending push notifications:

```
// src/index.html

<script>
  if ('serviceWorker' in navigator) {
    navigator.serviceWorker.register('service-worker.js')
      .then((reg) => {
        console.log('service worker installed');
        reg.pushManager.subscribe({
            userVisibleOnly: true // Always show notification when received
        }).then((sub) => {
            console.log(sub);
            localStorage.setItem('subscription', sub);
            let subKey = sub.endpoint.substring(40);
            console.log('endpoint:', subKey);
            localStorage.setItem('anonKey', subKey);
        });
      })
      .catch(err => console.log('Error', err));
  }
</script>
```

You can see here that after successful registration we subscribe `push notification` from `pushManager`. This will generate the subscription endpoint, which is responsible for sending notifications to all the users. The next step is to set up our `manifest.json` file:

```
{
"name": "Currency Converter PWA",
"short_name": "Currency",
"start_url": "index.html",
"display": "standalone",
"icons": [{
"src": "assets/logo.png",
"sizes": "512x512",
"type": "image/png"
}],
"background_color": "#db0000",
theme_color": "#000000",
"gcm_sender_id": "124917710568"
}
```

We have discussed all the properties in the last section, here we will add our `gcm_sender_id` for user notifications. You will get the `gcm_sender_id` from Firebase settings in cloud messaging. From there copy the server key as later on it will be required to send a web push notification.

Service workers are the most important for progressive web applications. Here is our default `service-worker.js`, we made some further changes according to our requirements. Ionic, by default, does not use the `sw-toolbox` library, which has reduced the code and efforts a lot:

```
// src/service-worker.js

'use strict';
importScripts('./build/sw-toolbox.js');

var version = 1;

self.toolbox.options.debug = true;

self.toolbox.options.cache = {
name: 'ionic-cache' + version
};

// pre-cache our key assets
self.toolbox.precache(
  [
'./build/main.js',
'./build/main.css',
'./build/polyfills.js',
'index.html',
'manifest.json',
'/'
]
);
```

We start with setting up the cache name, which we name with the version number specified. So whenever there is some new change we change the version and a new `cacheStorage` is generated. Next we cache all our static resources here using the `precache()` method and pass all static files paths. You will see that we have added another resource, which is not in the default file, as our application serve on browser at `localhost:8100` which in case went offline after first time will not cache it and you will see an error. If you don't want to add this just refresh the page twice as that will cache this request also.

Now that we have cache static resources, let's see how we have used caching strategies for dynamic content, such as API calls. We have two types of APIs used in our application on `HomePage` and `ListPage`, so accordingly we want different strategies for caching them. Everytime if we fetch the entire list we used `networkFirst` approach and for single currency rate we use `cacheFirst` approach. As both requests have a different API path we have applied conditions according to those paths. Also, we have set up different `cacheStorage` for these requests named `fixerio`, which will have all requests from `fixer.io` as its origin:

```
self.toolbox.router.get('/(.*)', function(request, values, options) {
if (!request.url.match("symbols")) {
console.log("===entire currency list API===");
return self.toolbox.networkFirst(request, values, options);
  } else {
    console.log("===specific conversion between two currencies===");
    return self.toolbox.cacheFirst(request, values, options);
}
}, {
cache: {
  name: 'fixerio',
  maxEntries: 10,
  maxAgeSeconds: 3600
},
  origin: /.fixer.io$/
});

// dynamically cache any other local assets
self.toolbox.router.any('/*', self.toolbox.cacheFirst, {
cache: {
maxAgeSeconds: 3600
  }
});

// for any other requests go to the network, cache,
// and then only use that cached resource if your user goes offline
self.toolbox.router.default = self.toolbox.networkFirst;
```

Also, we have cached any other local assets as `cacheFirst` and for other than that any request it will be `networkFirst`, which will go to network first and if offline then will get from cache:

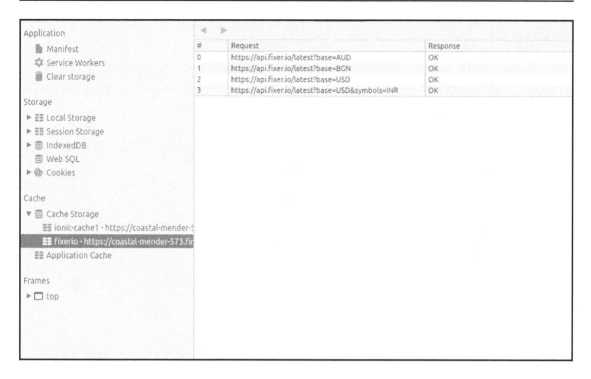

You can easily debug `cacheStorage` and `indexedDB` data in your Chrome developer tools.

The next important thing that we have to do is clearing old cache in case new versions of service workers create new `cacheStorage`. So we execute this during our activation of service workers:

```
    // Deleting old caches
self.addEventListener('activate', function(event) {
  var expectedCacheNames =
Object.keys(self.toolbox.options.cache.name).map(function(key) {
    return self.toolbox.options.cache.name[key];
  });
  event.waitUntil(
    caches.keys().then(function(cacheNames) {
      return Promise.all(
        cacheNames.map(function(cacheName) {
          if (expectedCacheNames.indexOf(cacheName) === -1) {
            console.log('Deleting out of date cache:', cacheName);
            return caches.delete(cacheName);
          }
        })
      );
    })
```

```
    );
});
```

We did some of the work for subscribing to push notifications, now we have to add a `EventListner`, which will show the notification as soon as it sends. We have the following:

```
self.addEventListener('push', (event) => {
console.log('Received a push message', event);

const title = 'Yay a message.';
const body = 'We have received a push notification.';
const icon = 'assets/logo.png';
const tag = 'simple-push-demo-notification-tag';

event.waitUntil(
self.registration.showNotification(title, {
  body: body,
  icon: icon,
  tag: tag
  })
);
});
```

We need to request for permission also, which we will asking in
`application.component.ts` initially. This will prompt a popup asking to send
notifications and as soon as you allow it you are now ready to receive push notifications:

```
// src/application/application.component.ts

Notification.requestPermission().then(function(result) {
if (result === 'denied') {
  console.log('Permission wasn't granted. Allow a retry.');
  return;
}
if (result === 'default') {
  console.log('The permission request was dismissed.');
  return;
}
});
```

For demonstration of web push notifications, we will be using curl requests, which need the
endpoint, which we can get from the `localstorage` as we saved it and the Firebase key
that we copied from the panel:

```
$ curl https://android.googleapis.com/gcm/send --header "Authorization:
key=AABBHRWs_ug:APA91bEiHIECzZLiMJAB-_icXkI7xMDNRha_-
LH5efZfUP7u0xAALa2eZfEzxDItC2qzRk0SjrLcgJZj2ZCTFBiMGRbUCM73K5-
O8xBz_ydAEtmwMaEb1gbxAbIOtsRi4Pn0UzItvCHp" --header "Content-Type:
application/json"  -d
'{"registration_ids":["d7BRrMPamaU:APA91bEsRCOFmt746cHPINBam7PRIW22Jp-
Ufuh_BK_1UfwdJmgMeONCn8yKWN6eskwK7EYi64bkZIruPPaCe6kWJsNhGnvnlZzETdV8WhsItW
IdXCBYJUnXdlmjJnOPRNgLlBMDb-vA"]}'
```

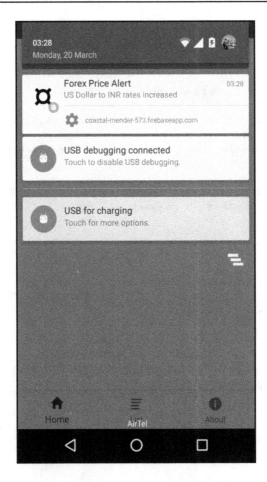

We have successfully sent a web push notification now and, next step from here would to to integrate this push API to integrate to your backend. We can also send custom payload in notifications, but for that we need to add encryption and make changes on the client side and server side. As it's not in the scope of the current context we are skipping this, but you can follow this Google documentation for more information (`https://developers.google.com/web/updates/2016/03/web-push-encryption`).

We are done with the major changes that we have to do in our PWA application, we now just have to start writing our code and integrate the exchange rate APIs. Let's look into our `CurrencyService`:

```
// src/providers/currency-service.ts

import { Injectable } from '@angular/core';
import { Http } from '@angular/http';
```

```
import 'rxjs/add/operator/map';
import { StorageService } from './storage-service';

@Injectable()
export class CurrencyService {

  constructor(public http: Http, public cache: StorageService) {
  }

  loadCurrencyList(base:String) {
   return new Promise (resolve => {
      let url = "https://api.fixer.io/latest?base=" + base;
      this.http.get(url).map(res => res.json()).subscribe(data => {
        this.cache.setObject('base_'+ base, data);
        resolve(data);
      }, error => {
        resolve("Data not found");
      })
   })
  }
```

For fetching the currency according to the base currency we used
the `loadCurrencyList()` method, which then calls the `fixer.io` API to return the results
and also cache it using service worker. We have filtered this type of request to be
`networkFirst`, so everytime it will fetch the latest data from the server. In case a device is
offline, it returns the last cached result and if in case the request was not cached it will go to
the error block and we will resolve it as `Data not found`. At the same time we are also
storing the data in `IndexedDB`:

```
convert(from,to) {
  return new Promise(resolve => {
    this.http.get("https://api.fixer.io/latest?base=" + from +
    "&symbols=" + to).map(res => res.json()).subscribe(data => {
      resolve(data);
      }, error => {
        this.cache.getObject('base_'+ from).then((value) => {
          if(value) {
            resolve(value);
          } else {
            resolve("Data not found in offline State");
          }
        })
    })
  })
  }
}
```

On our `HomePage` we fetch the one-to-one currency conversion rate and for that we use the `convert()` method. Here our approach is `cacheFirst` and we will return the `cacheStorage` data first and if it is not found we will call the `fixer.io` API. In case we didn't cache a specific request and we are offline then it will fallback to error block, where we will first check if we have the base currency list already stored in `IndexedDB` while we fetched its entire list. If we have that in our storage then we can easily find the second currency in the list.

Let's look into our `ListPage` class from where we will be calling the `CurrencyService`:

```
// src/pages/list/list.ts

import { Component } from '@angular/core';
import { NavController, NavParams } from 'ionic-angular';
import { CurrencyService } from '../../providers/currency-service';

@Component({
  selector: 'page-list',
  templateUrl: 'list.html'
})
export class ListPage {

  baseCurrency:any = "USD";
  currencyList:any;
  queryText:'';

  constructor(
   public navCtrl: NavController,
   public navParams: NavParams,
   public currency: CurrencyService) {

   currency.loadCurrencyList(this.baseCurrency).then(val => {
      this.processData(val)
   });

  }

  processData(data:any) {
   console.log(data);
   let rates = data.rates;
   let arr = [];
   for (var key in rates) {
        if (rates.hasOwnProperty(key)) {
            arr.push({'name':key, 'price':rates[key]});
        }
    };
    this.currencyList = arr;
```

```
    }

    setBaseCurrency() {
      this.currency.loadCurrencyList(this.baseCurrency).then(val => {
        if(typeof val == "object") {
          this.processData(val)
        } else {
          alert(val);
        }
      });
    }
  }
```

We set the `baseCurrency` initially to `USD` and according to how a user selects the base currency we fetch the data calling the `loadCurrencyList()` method of `CurrencyService`. We have discussed most of almost important points and code snippets of our progressive web application. You can find the entire codebase with the chapter and play around with it. We know that our application will be fully functional only if we host it on HTTPS, so I have deployed the www folder with Firebase hosting:

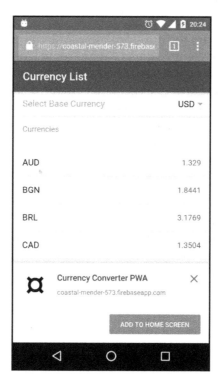

You can find the application hosted at `https://coastal-mender-573.firebaseapplicati on.com/` and as you start to use the application frequently you will find a promt asking you to add this to your mobile home screen. After you add it to your **home screen** you will see an icon placed on your home screen and as you click on that it will open a separate fill screen and it will appear like a native application. I must say that you should all be excited to see this and you should try building a PWA application of your own now, which will strengthen your skills further as practically developing an application will give you an upper hand rather than just reading the concepts.

The future of PWAs

Progressive web applications gives us so many possibilities such as the ability to install, send push, and offline support, which bridges the gap between mobile web and native applications. Currently, the problem is that iOS and Safari do not have the support for service workers, which is still under consideration. There are assumptions going on that in the latter half of 2017, they might come up with service worker support, which will help enable PWA features.

Progressive web applications are the next step forward in improving mobile web. With growing mobile traffic day by day this make a lot of sense. Also, from last some time we were seeing that many websites are forcing their users to use mobile web to install native applications, which is really annoying. One of the famous e-commerce giants in India, Flipkart, even discontinued its mobile web version once with a banner that asks users to install their native application. Later, as they realized this mistake, they were the first ones in the industry to support PWAs by releasing Flipkart lite version on mobile web.

Statistics shows that in the US smartphone application downloads decreased to 20% in the past year, which shows that the market has become saturated and consumer needs are changing. You can find a lot of case studies here (`https://developers.google.com/web/s howcase/2016/flipkart`), which have key insights how into PWAs have positively affected product growth and user engagement. There are many PWAs such as MakeMyTrip, Housing, and so on, you will found, which helped me uninstall native application to free my device space and use PWAs version.

Summary

I would personally say that PWAs support for Ionic and discussing this topic in this last chapter is the best approach to wrap up our exciting journey. You have already understood now the power of a PWA, and which you can extend to your normal web applications also. Also, we can't say here that PWA will replace native applications as there are low-level APIs that you only have while you are building your application with Cordova and using Cordova plugins.

From recent blog posts from the Ionic team, it looks like we will be having desktop support with electron coming really soon. I would recommend every user or developer to the subscribe to Ionic blog, which is the best way to be up to date about what's going on with Ionic. I have tried my very best to make sure that the content of every chapter is up to date, but you can see how actively the project is growing, so make sure that you put the effort in to also be
up to date. The Ionic forum is another useful resource that has been helping me for the last three years in my journey to learn Ionic. Most of the time you will find that the problems you are facing have been already asked in the forums and instantly you have your answers. Also, you can ask questions anytime, but make sure that you have done your homework before that otherwise there can be chances that your question went unanswered. By homework I mean and search existing discussion in forums, check for issues in GitHub as many times it can be an issue that is already marked and you don't want to waste your time. Most of the time we might have some issues that are related to Angular and are not specific to Ionic components, or APIs. So for that try searching on Google using the keyword Angular rather than Ionic as that will yield more appropriate results.

Index

physical Web
  about 185, 188
  URL 189
pipe 8
playground
  about 138, 140
  URL 140
Plugman 74
progressive web applications (PWA)
  about 201
  advantages 202
  case studies, URL 222
  disadvantages 203
  future 222
  manifest file, generating 205
  native application experience 204
  need for 201
  service workers, using 207
  support in Ionic 203
  with Ionic 204
providers 8
proximity based screen lock
  creating, with BLE 190
proxy server
  setting up, for Ionic app 147
pull-to-refresh 61, 64

S

SASS
  customizing 17
  theming up 16
security
  about 149
  security points 153
  sensitive data outside app, extracting 151
  SSL, enabling 150
  storage, securing 152
  whitelisting 149
segments
  versus tabs 41, 43, 44
service worker toolbox (sw-toolbox)
  about 210
  URL 210
service workers
  URL 207

  using 207
slides component
  about 35, 36
  URL 37
social sharing plugin 85
splash screen plugin 78
splash screen resources
  building, for platform specific resources 19
  creating 18
streaming plugin 98

T

tabs
  URL 44
  versus segments 41, 43, 44
TasteBite app
  building 166
  creating 165
  CRUD operations, performing 173
  data, securing 178
  data, securing with AngularFire2 172
  data, structuring 178
text to speech plugin 89
toast component 39, 40
token based authentication
  about 144
  benefits 144
  Cross Origin Resource Sharing (CORS) 146
  Cross Site Request Forgery (CSRF) 148
  proxy server, setting up 147
tools
  about 135
  Ionic creator 138
  Ionic View 135, 136, 138
transfer plugin 98
TypeScript
  about 8
  features 13

U

Universally Unique Identifier (UUID) 78

V

virtual scroll 64
vPlanet Commerce

www.ingramcontent.com/pod-product-compliance
Lightning Source LLC
Chambersburg PA
CBHW082118070326
40690CB00049B/3609